Praise for the I.T. Sales Boot Camp seminar:

"The I.T. Sales Boot Camp shows you how to get to the decision-maker and address their issues. It demonstrates how to close SIGNIFICANT sales."
—Chis Palen, Vice President, Dunn & Bradstreet

"Talk about positive feedback! My team is raving about the I.T. Sales Boot Camp! It has a solutions-based methodology, proven strategies for effective penetration in an ever changing marketplace. Thank you!"
—Ed Abner, Vice President of Sales and Marketing,
Sysorex Information Systems

"I.T. Sales Boot Camp created energy and enthusiasm that has lasted well beyond our three-day meeting. Thanks for your creativity. . . ."
—Brad DuPont, Director of Sales, Jetform

"Your program left a powerful impact on our sales team! We enjoyed the upbeat, no-holds-barred style. Brian is more a partner than an author."
—Robyn Gross, Manager, American Mobile Satellite

"Brian has the pulse on selling. More than that, he is in tune with how people WANT to be sold. This is a GREAT CLASS: in fact, the best I have ever attended."
—Frank Olivera, Vice President of Sales, ECE

"The information from I.T. Sales Boot Camp was extremely beneficial because it can be applied immediately on the job."
—Jennifer Schaus, Account Consultant, Dunn & Bradstreet

"I have been selling for 26 years and I came away from the I.T. Sales Boot Camp with some valuable information and techniques."
—Robert Musitano, Relationship Manager, Dunn & Bradstreet

"The I.T. Sales Boot Camp is the best sales training I ever had!"
—Suzanne Davids, Account Manager, SGData

"The I.T. Sales Boot Camp is an excellent framework for successful selling in any complex environment."
—Gene Murray, Senior Sales Representative, Group 1 Software

"I.T. Sales Boot Camp provided me with effective hands-on practical skills that I can practice and implement in my client contacts on a daily basis. I highly recommend it. . . ."
—Vince Defrancisci, Director of Sales & Marketing,

Computer Analytics"Brian Giese clearly knows the techniques for making a person understand what makes someone buy something."
—Ben Schmitt, Sales Representative, First USA Technology

"I.T. Sales Boot Camp takes the mystery out of the sales process."
—Dave Teti, Senior Account Executive, Siebel Software

"Brian Giese's history in the IT field is very valuable. . . . "
—Preston Hall, Client Account Manager, Process Consulting

"I've attended 20+ sales seminars and workshops. The I.T. Sales Boot Camp is the most valuable workshop I have ever attended!"
—Joel Burnett, Program Manager, Rackspace

"No matter what your sales experience level, you are assured to get ideas and be able to formulate a sales plan for your accounts."
—Jack Harrian, Vice President of Sales, Radiant Systems

"Don't talk to potential clients with having completed the I.T. Sales Boot Camp!"
—Kathy Finnegan, Business Development Consultant, NUA

"Sign on ASAP!"
—Howard Abroms, Vice President, Smith Hanley Associates

"I gained a great deal of confidence and am excited to face new sales challenges as I feel well prepared to meet them."
—Brett Martinez, Technical Manager, Lockheed Martin

"I.T. Sales Boot Camp is very beneficial to me as a technical person moving into the sales field. . . . ."
—Chris Kyhl, Technical Recruiting Manager, Quality Consulting

"If you are looking for ways to get out of the same old path and are looking for different and proven ways to improve your sales skills take a look at the I.T. Sales Boot Camp."
—Nate Evans, Account Manager, AAA Networks

"I.T. Sales Boot Camp packages concepts which are next to impossible to learn by yourself. . . . this is a great tool for better selling!"
—Henry Chmielnicki, President, Profit Recovery Systems

"I strongly recommend the I.T. Sales Boot Camp to even hard-core salespeople as it gives a lot of in-depth insights and a new perspective to the sales process."
—Raghu K. Rao, Director of Software Services,
Kanrad Technologies

"The Boot Camp allows me to sharpen my skills more than the other sales courses I've taken—COMBINED!"
—Ro Chapple, Account Manager, Bell South

I.T. SALES BOOT CAMP

Sure-fire Techniques for Selling Technology
Products to Mainstream Companies

★ ★ ★ ★ ★

By Brian Giese

Adams Media Corporation
Avon, Massachusetts

To my wife, Patricia

Published by Adams Media Corporation
57 Littlefield Street, Avon, MA 02322. U.S.A.
www.adamsmedia.com

ISBN: 1-58062-538-X

Printed in Canada

J I H G F E D C B

Library of Congress Cataloging-in-Publication Data
Giese, Brian.
The I.T. sales boot camp : sure-fire techniques for selling
technology products to mainstream companies / Brian Giese.
p.   cm.
ISBN 1-58062-538-X
1. Telecommunication equipment industry. 2. Information technology--
Marketing--Handbooks, manuals, etc. 3. Telecommunication--Equipment
and supplies--Marketing--Handbooks, manuals, etc. I. Title.
HD9696.T442 G54   2002
004'.068'8--dc21   2001055207

This publication is designed to provide accurate and authoritative information with regard to the subject matter covered. It is sold with the understanding that the publisher is not engaged in rendering legal, accounting, or other professional advice. If legal advice or other expert assistance is required, the services of a competent professional person should be sought.
— From a *Declaration of Principles* jointly adopted by a
Committee of the American Bar Association
and a Committee of Publishers and Associations

Cover illustration by Stephen Marchesi.

This book is available for quantity discounts for bulk purchases.
For information call 1-800-872-5627.

Contents

Chapter Eight
Engaging the Enemy:

Chapter Nine
Managing Your Buying Allies:

Chapter Ten
Coordinating Your Sales Channels:

Appendices

Acknowledgments

I HAVE MANY PEOPLE AND COMPANIES to acknowledge for their inspiration and important contributions to the writing of this book:

To my clients, those buyers who have implemented the *I.T. Sales Boot Camp* methodology inside their companies or those individuals who have taken the material and used it to shorten their sales process, I thank you. To the thousands of graduates of the program who refined and improved the process and contributed to the success of those who followed, I applaud your tenacity and ingenuity.

This book would not have been possible without the contributions of many family, friends, sellers, and buyers:

To my wife Patricia, who made countless emotional and professional sacrifices and who continues to inspire me in all my efforts. To my children, Billy, Richie, and Kathy, and my mother Anita, whose inspiration over the years has made this work possible.

To my associates, Diane Ferguson, for her unflinching loyalty and ability to translate buyer needs, and Lamont Wood, who guided me through the complex business world of publishing.

To Myron Radio, my good friend and trusted champion, for his wit and wise advice. To Marc Derrickson, an IT sales best-practice expert, who helped with editing and overall design over this long journey and who was there every step of the way.

To the buyers, coworkers, champions, and friends who have, over the years, been significant contributors to my learning. They are responsible for the stories contained in this book and, ultimately, for its success. While it is impossible to acknowledge all of them, some of them include (in alphabetical order) Ed Abner, Michael Bermel, John Cook, Sandra Crowe, Bruce Deming, Roy Gemberling, Norman Gilfand, Tom Hamel, Jim Holder, Tim Karney, Jim Linton, Ted Milkovich, Alfred and Barbara Navarro, Marie O'Brien, Dennis "DOC" O'Connell, Roland Schumann, Marc Wallace, Marc Weber, Don Wilson, and John Wyatt.

And finally to the knowledgeable people at the Strategic Account Management Association for their valuable contribution, and to Jeff Herman, my literary agent, for identifying this opportunity and mentoring its success.

Introduction

THIS BOOK IS WRITTEN for information technology sellers. Its aim is to provide new and experienced sellers alike the current tools, tactics, and strategies that, used together, will dramatically shorten the process of selling complex solutions to large accounts. Tapping into years of research, evaluation of best-practice sellers, and real-life experience, I have tried to provide the seller with a methodology for targeting, closing, and managing IT sales opportunities.

I.T. Sales Boot Camp is not merely a book on sales techniques. For one thing, it is based on the time-tested seminar of the same name, which is used by world-class sales teams around the world. For another, it is for those sellers, both new and experienced, who are serious about providing technology solutions to large accounts and who are looking to shorten the sales process. Like the buyers to whom they pitch, IT sellers are forward-thinking learners who need professional, specialized skills to do a superior job. To them, *I.T. Sales Boot Camp* offers a wealth of tactics and strategies that, I believe, constitute a new selling paradigm.

Of all the sales fields, information technology (IT) is the fastest growing. According to the U.S. Census Bureau, approximately seven million people worldwide are involved in selling IT solutions and the number is growing by an astonishing 60 percent per year. That growth reflects a proliferation in the number of new and growing software, hardware, Internet, and service firms. Not surprisingly, many of these newly minted IT salespeople migrate from one of the more traditional sales arenas: finance, retail, insurance, healthcare, or real estate. Very soon they discover the painful fact that informs every page of this book: Selling technology solutions to large buyers is a unique endeavor.

Why is this so? I will offer five reasons:

Number One: Technology-oriented buyers have vision and smarts. In fact, in many situations, the buyer will know more about the product you represent than you do. They are insatiable learners, usually well educated, and given a little experience, they are not afraid to show off their knowledge. If you are used to getting by on limited product knowledge, a snappy sales script, and an innate ability to read people, then you are about to enter a very different, and sometimes frustrating, world.

Number Two: The IT business is known for its intimidating language. It is full of specialized acronyms, phrases, and meanings. If you speak the language, you will know the right word for the right occasion. You will know the difference between a "megabyte" and a "megabit," between "Customer Relationship Management" and "Sales Force Automation." If you do not speak the language, you are at an immediate disadvantage. Technical lingo is so important to the IT buyer that it is doubtful you will be able to credibly carry out the simplest job-related conversation without being equipped with it.

It is not that IT folks like to build and protect a secret club. Think of the Inuit in the Arctic, whose language is said to contain a vast number of words to describe snow, ice, and related weather conditions. It is understandable: Snow and ice define their environment and passing along precise and subtle descriptions can be a matter of life and death. Likewise, IT buyers use language to reliably define a complex, ever-changing world. As a result, a serious IT seller has to invest in learning the language. It is a never-ending investment, since new technologies—each with an accompanying load of new technical terms—can appear, flower, pass into obsolescence, and be forgotten in the course of a couple of years. While this book will not give you comprehensive cutting-edge lingo and techno jargon, it will offer enough background and glossaries to make you literate in IT selling.

Number Three: An IT seller is a global person with global connections. This is an industry that reorients the phrase, "Think globally, act locally." In this world, you have to think *and* act globally.

That has a profound implication: It means that selling technology must be a team sport. While in other, less complicated, transaction-oriented sales you may be able to swing an entire deal solo, when selling solutions you always have to bring in others. The more complex the solution and the higher the revenue volume, the more team-oriented the sale. That means you will have to rely on product managers, pre-sales technical support, bid or proposal writers, sales management staff, subject matter experts—a cavalcade of people—who do not report to you. As you will read frequently in this book, best-practice IT sellers do not see themselves as salespeople but as team captains who facilitate the selling process.

Another feature of globalization is that business alliances are often made between and among people who have never met. Sales territories are so fluid that you will probably find yourself thinking in terms of time zones rather than street names. But that does not mean you will be jetting to Paris and Bangkok at the drop of an expense account voucher. On the contrary, you will be doing more and more selling from your desk, because what globalization involves today is *using* technology to sell technology.

In future chapters, I will describe the team selling approach and its impact on the buyer. With the help of quizzes and worksheets, we will look at all of the areas of your team in detail and how to deploy the team effectively.

Number Four: The competitive landscape changes at an alarming rate. You may be familiar with Moore's Law, the observation that the power of computer chips doubles every eighteen months, while staying the same price. This makes last year's business plans as obsolete as 5.25-inch floppy disks and as useful as buggy whips. Technologically adept competitors who can change at a moment's notice are searching for new markets to exploit, and have radically changed the competitive arena. At any moment you can find yourself under devastating attack.

Anticipating competitive change requires you to reconsider your definition of a "competitor." No longer can you afford to focus only on the players in your industry who produce similar products and who prospect for the same kinds of customers. Today's sellers expand the definition of "competitor" to include "any company that is capable of meeting your buyers' needs as well as you can, or better." In this book we will examine, in detail, how IT competition works, what to look for, and how to launch a synchronized counterattack.

Number Five: In any complex, solution-oriented sale, each buyer measures value in a different way. In fact, the same type of buyers—ones that share similar responsibilities—will measure the value of one solution differently at any given time. With two similar buyers in the same marketplace, one may place significant value on a particular solution while the other may see the same solution as actually detrimental. This represents the biggest challenge in providing an IT solution.

As a best-practice IT seller, your answer to this challenge is to identify and link value to the buyer. This can be done using what I call the SECRET weapon. As you will learn in future chapters, the SECRET weapon represents my realization that there are six areas of value that a technology solution can affect (one for each letter of the word "secret"). The buyer may attach value to one or all of these "buckets" of wants. Some are measurable and some not, but all can be significant reasons for the buyer to move ahead in the sale.

But it is not enough to identify value—you, the seller, must then link the value to the solution, a crucial step that the seller often skips. Throughout the book, I will discuss how to link value at an emotional level, giving the buyer a significant reason to buy—from *you*—while shortening the sales process.

The *I.T. Sales Boot Camp* Method

Traditional sales approaches describe the sales process as a "funnel." In this funnel are poured large amounts of prospects in the hope that one will come out the other end, resulting in what one would hope would be a successful sale. Many readers understand the funnel and are trained in this approach; I see it all the time in my consulting practice. Sellers cold-call to create buyer demand. They probe to find the "right" person to talk to or the right value proposition. They repeat the same process over and over again, usually with no positive results.

After interviewing sellers and their managers and picking apart this methodology, I was struck by the fact that complex and specifically technological sales are different. The things that happen in the IT selling process are very political—timing is critical and a team approach is mandatory.

I.T. Sales Boot Camp cuts to the chase about what really goes on in this selling arena. This method works, it's time-tested and true. It is the product of thousands of interviews, numerous studies, and years of experience, and is now being used by world-class IT sales teams across the world.

As you read this book, keep in mind that it is based on a "process" and uses a "methodology." As with any process, it is structured as a series of steps, one logically following the next. I encourage you to read the chapters in sequence and to use the information you gain from one chapter to build upon the next. It might not be easy work, hence the title: *I.T. Sales Boot Camp*. The following chapters will put you through your paces, maybe even force you to rethink some ideas about the best ways of selling technology solutions. But you'll emerge from this boot camp fitter for the effort.

I hope you enjoy the book as much as I have enjoyed writing it.

Brian Giese
October 2001

Chapter One
Mapping the Battleground: How to Spot Your Stakeholders

NOT LONG AGO I met with a salesperson from a new software company. He was launching what top management regarded as a "clear sales plan." Unfortunately, *he* didn't see it that way. He confided his frustrations with trying to create demand for a new, expensive, but potentially "disruptive" solution. Disruptive in the sense that his solution would be replacing old legacy systems with new ones at a significantly lower cost. Unfortunately, buyers were just not biting. Lamenting the long and drawn out deal-making cycle, the salesperson then asked the famous question: "How do I shorten the sales process?"

Shortening the sales process—by as much as 50 percent—is exactly what this book is about. In these pages, you will discover new sales strategies for landing large accounts as well as tactics for approaching these potentially lucrative target buyers. For experienced sellers, you may even experience a shift in the way you think of and handle your accounts, or a change in attitude about them.

To understand the scope of the challenge, let's look at the kind of selling that is the polar opposite of IT selling.

The Dawn of a New "Funnel"

Twenty years ago, when I started in sales, I was paying my way through college by being a salesperson on the floor of a large furniture store in Washington, D.C. "Out the door, see no more!" was the battlecry for the sales staff. We assumed that if the buyer left without buying anything, he would never come back, so we had to close the sale during the buyer's first, and possibly only, visit. In such an environment, you needed to understand individual buying behavior and possess finely tuned closing skills.

For forty years and running, "Out the door, see no more!" has worked well for this particular furniture retailer. It is a storefront business, where the sales process is direct and short, sellers are dealing with few buying influences, and results are measured by the number of successful transactions. We considered a complex sale as one where both husband and wife were in the store together and you had to sell to both of them.

I later learned that traditional sales approaches such as this one are described as a "funnel" into which are poured large amounts of prospects, with the hope that one will come out the other end (see **Figure 1-1**). In the best of all worlds, the funnel would be shallow, leading to a quick and successful sale. This philosophy is known as, "Mile wide, inch deep."

Soon after college, I began selling technology solutions to large accounts, where the sales process truly is complex. Whom should I call first? Who is the ultimate decision-maker? What are the politics within the account? Who has the most influence and with whom?

There are political dynamics in every organization, both healthy and unhealthy. Healthy politics are aimed at the best interests of the company; they reduce costs or increase revenues. Unhealthy politics are those that are centered on the promotion or advancement of individuals, on how to gain advantage in some internal competition. While such an environment is negative, the seller should work within it, accept it for what it is, and focus on getting the buying decisions that the seller wants.

Figure 1-1: Traditional Sales Funnel

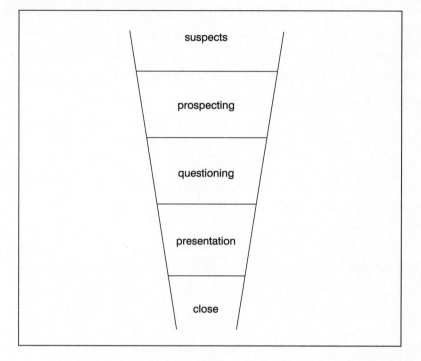

I quickly discovered that in technology sales, more than in other areas, a great number of people can influence the process. All the twists and turns that occur as events of the sale unfold make the process very political; timing is critical and a team approach to winning business is mandatory. In such an environment, the traditional approach of "mile wide, inch deep" simply does not work.

Looking at the steps involved in a complex technology sale, I found that the funnel is actually *inverted* (see **Figure 1-2**). In fact, a more realistic and successful philosophy should be, "Inch wide, mile deep." This new thinking is counter to the way many salespeople have been trained. It is upside-down and inside-out, with more steps and added complexity. Noticeably, successful sales organizations spend much of their time planning which accounts or markets to approach and how to approach them. Many accounts are sifted before entering the funnel based on the company's selling capabilities, cost structure, or other related business reasons. Putting less—but better qualified—buyers in the funnel increases buyer satisfaction and reduces the cost of sales, which ultimately increases profitability.

Figure 1-2: IT Sales Funnel

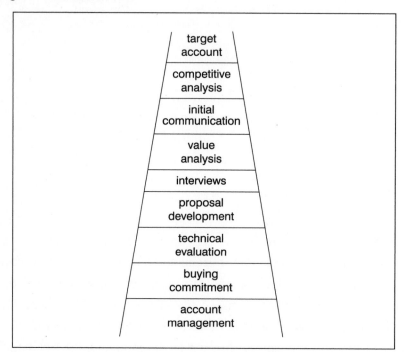

One look at this IT sales funnel should clearly illustrate just how complex the selling process is and how important it is to follow a system and methodologies, such as those offered in this book. Notice, for example, how, when selling solutions in a corporate or government environment, the outcome involves groups of people. Committees or boards of review decide on large purchases, making the traditional funnel obsolete. You, as an IT salesperson, need state-of-the-art political skills to compete in this environment.

Today's IT buyers are also looking for more value from their vendors and are demanding more sophisticated selling strategies. More value in a business-to-business sale means the salesperson is not just pitching a single product, but is also offering a suite of bundled hardware, software, and professional services—and offering them as "solutions" to solve business problems. No single salesperson can understand and tie together all of these value components. *Salespeople*, working in coordinated selling teams, can.

Beating the Beta Blues

To make matters even more interesting, complexity in the buyer's world is only half the story. The technology wizards in your own company are probably trapping you between a rock and a hard place. Let's face it: The harsh reality in modern technology companies is that most solutions are released to the market long before they are ready.

Version 1.0 commonly does not work properly and sometimes never works in the configuration that was originally advertised. Software has bugs, hardware is difficult to configure or simply doesn't work as promised, and client service is often not yet in place.

Given these limitations, should you wait until a solution is technically "proven" before you begin to sell it? Unfortunately, most organizations cannot afford that luxury. In fact, you face a sobering paradox shared by all your competitors: The initial sales of a product or service are expected to fuel further development of that solution. The initial buyers are the beta testers, or early adopters, who eagerly await version 2.0, even as the reality of version 1.0 is still sinking in.

When Lotus first delivered Notes in the early 1990s, the group software product required significant customization. Large end-user organizations understood this and funded the growth and development of the software, allowing for a large market to grow and mature.

Every company that delivers new technology to market has a similar story. The question is how well can the initial sales effort propel the company through the start-up and early development phases before delivering a durable, working solution to market. After arriving at a mature solution that works as advertised and that can be readily duplicated, the technology can be scaled to the mainstream market. At this point, the salespeople behind it will be in a position to make money.

To achieve this objective, the seller has to shorten the sales process or the solution will die on the vine. So, in response to the question posed by the software salesperson described at the beginning of this chapter about shortening the sales process, I ask, "What is the single, obvious benefit your product provides to improve or replace an existing expense that is already in the buyer's budget?"

In other words, how does this solution save or make money for the buyer? Once this code is cracked, the sales process will shorten significantly, because the motivated target buyer will start coming into focus.

The IT Seller in Action!

Before enlisting in I.T. Sales Boot Camp, let's look at a day in the life of a best-practice IT seller who represents a supplier of wireless palmtop solutions to large accounts.

It's Wednesday at 2:00 P.M. Allison, the seller, meets with her sales team to discuss the prospect and the team's strategy. As a team, they participate in a SWOT (Strengths, Weaknesses, Opportunities, Threats) analysis of the potential buyer—its internal strengths and weaknesses, and external opportunities and threats. The team comes up with action items to advance the sale and then modifies the presentation.

Wednesday, 9:00 P.M. Allison returns to her Chicago hotel room after a dinner meeting with a representative of another company. Her colleague has given her an important tip: The blue-chip buyer he is wooing tomorrow will want to know how her company's solution will help its sales force. Allison plugs in her laptop and fires off an e-mail to a product manager in Asia, where it is 11:00 A.M.: "Please send latest slides on sales force automation solutions."

Next, she logs onto the company intranet, which contains a mine of information for its global sales force. She can download white papers, join live problem-solving chats, or view slides of products that are still in alpha stage. Tonight, she'll grab the latest research on the growing market for sales force automation tools.

Thursday, 8:00 A.M. Before breakfast, Allison visits Web sites that she had earlier bookmarked—including CNN, *Wall Street Journal*, Hoovers Online, and competitors—to find data that will help customize her presentation. *VARbusiness Magazine* is reporting fractures in the partnership between two of her competitors. She checks the report against other sources and decides that the news is reliable. Now she has ammunition against her largest competitor: "They don't have a future strategy because their current strategy is on the rocks."

Thursday, 11:00 A.M. Allison arrives at the prospective buyer's office and proceeds with a simple, low-tech presentation—no hardware except for her palmtop. She chooses not to use a laptop for the presentation because she wants to be spontaneous in this small group setting.

When the Chief Information Officer asks Allison what type of connection is required to transfer e-mail from palmtops to desktop computers, she is momentarily stumped. But she composes e-mail on her palmtop (which is equipped with a wireless modem) and sends an urgent query to her Proposal Leader in California. The e-mail exchange adds value to his presentation: Instead of offering an artificial demonstration, she shows how the technology has become an essential tool in her business life.

Thursday, 4:00 P.M. As soon as her final call wraps up, Allison logs back onto the Internet and replies to questions she couldn't answer in one meeting, laying the groundwork for a chance to make another pitch.

Drawing a MAP

The experience of our fictitious seller illustrates several keys to successful IT selling. The most obvious is the importance of organization, speed, and momentum. Smart IT sellers leverage technology to steer clear of speed bumps that can slow a sale. And using technology to stay close to the buyer helps the seller deliver a personal brand message: I'm fast, I'm connected, and you can depend on me.

But there is a less obvious, even more important message: You have to know your prospective buyer's business—its internal strengths and external challenges—and how you can add value to his enterprise. How do you accomplish this? How can you develop a thorough business understanding of your buyers and be able to follow your intuition about new sales opportunities?

A good way to start is to list specific criteria that describe current buyers, including size and type of business, revenues, markets they serve, market strength, and buying power. Then, study your list for ideas on where to look for similarly qualified prospects. Ask your current buyers for referrals and information about their associates that might provide you with more leads.

Salespeople, planning accounts on a national scale, should extend their search to one or more major industrial directories. Focus on product or service categories similar to those of your existing buyer base. Sellers with a very diverse buyer base can select a few key buyers, target their markets, and research their competitors to either target them as prospects or uncover new applications for their products.

Before you begin your selling efforts, create and fill out a Major Account Profile (MAP) form to use as a guideline to search for new opportunities (see **Figure 1-3**). The form contains spaces for all pertinent information on your account: company name, address and telephone number, key contact names and titles, product or service description, and definition of primary needs. It has extra space for notes, and it is easy to revise and expand. Included is a ranking system to help you assess each prospect's potential quickly and easily.

Figure 1-3: Major Account Profile

BUYER:		
Project / Opportunity Name:		
Description of Opportunity: *(Include customer background, business wants, technical wants and Your Solution.)*		
Estimate of Opportunity Size / timing:	$ initially / when?	Total $ potential / timeframe
What is the overall strategy to address this Opportunity:		
Major Roadblocks / Showstoppers / Issues:		
SELLER:		
Other critical internal team members involved in opportunity:		

Name	Phone	Role

Date Of Most Recent Update:

Major Progress Events Since Last Update:

Tactics / Plans / Next Steps

I know what you're thinking: another form to fill out. Thirty-, sixty-, ninety-day forecasts, activity planners, cold call sheets, contact managers, to-do checklists, and on and on. Although these forms are important for keeping track of day-to-day account information, there is always a "good" reason not to get them done. But suppose you could get a grip on the politics of an account with a minimum of paperwork? Suppose you knew exactly where every one of your potential buyers was at every stage in your sales process? That information would be worth putting down on a form.

Your MAP profile can only begin when you have identified the players inside the account. This may seem obvious but it is a critical step for success. Yet most salespeople skip this entirely and fail to understand the political dynamics that are rampant in large accounts.

In the MAP profile, we refer to the players inside the account as *stakeholders*—those people who influence the buying decision. Within the account itself, there will be at least five types of stakeholders as shown in **Figure 1-4:**

- Financial Stakeholder
- Champion Stakeholder
- Technical Evaluator Stakeholder
- Cost Evaluator Stakeholder
- End User Stakeholder

Sounds like politics. The fact is, experienced sellers are very sensitive to the politics within their accounts. They are like long-tailed cats in a room full of rocking chairs. They are alert and sensitive to interaction among the stakeholders. They continually watch, in advance of the sale, and gather information about the power and the influence of the stakeholders.

Whether the politics in the buying organization are healthy or unhealthy, politics cannot be ignored. One stakeholder alone could wreck the deal if he or she is unhappy about it. Often the smaller the transaction, the more closely coupled the stakeholders become, and there are times when one stakeholder can wear many hats.

Figure 1-4: Stakeholder Wheel

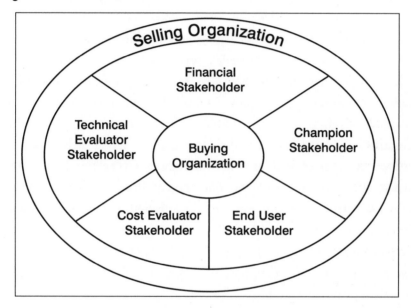

Most sellers, both new and experienced, have a difficult time with account politics. But when the chase is on for the sale, connected sellers—ones that have contacts with multiple levels—usually win the day.

Entering the account at only one of these levels is not enough, no matter how promising the initial reception. All too often, salespeople will enter an account at one level—and stay there. But if that stakeholder leaves the company, switches positions, or loses political power, the seller is in trouble. The MAP form and the Stakeholder Wheel will help the seller penetrate the account at many different levels, providing the seller and the sales team with valuable information for the attack.

To better understand internal politics and their effect on sales, let's take a look at the characteristics of each type of stakeholder.

Financial Stakeholder

Financial Stakeholders are those who have the authority to use funds on a discretionary basis and, should they find value in your solution, can transfer funds from other sources to make the procurement; they decide to have the check written. Flanked by a team of other evaluators, Financial Stakeholders provide final judgments based on the company's values and vision and how your solution fits in. In smaller companies, Financial Stakeholders typically have a title such as "CXO" (Chief "X" Officer, such as Chief Information Officer); in larger companies, there may be a group or committee fulfilling that role.

Modern CXOs are particularly numbers-oriented. They are directly involved in quantifiable business objectives, such as project and product management, sustaining competitive advantage, and gaining market share. You, the seller, need to show what benefits will arise if they buy your solution.

You probably won't have a lot of time to make your pitch. Time is in short supply for Financial Stakeholders. "Do you have time to talk?" is probably an unwise opener, unless you perceive that the stakeholder immediately sees some real, up-front value. Otherwise, you will probably be passed off to a technical staff member who can evaluate your offering.

By showing respect for their time, you are showing respect for these executives on a personal level. Financial Stakeholders are usually more comfortable cutting to the chase. In the end, they will value the respect you have shown for their time.

Early in my sales career I learned this lesson the hard way! I was lucky enough to call on the Financial Stakeholder of a very important and valued account. The American Red Cross is the largest not-for-profit organization in the world and was interested in upgrading its entire computer infrastructure. The Red Cross's Financial Stakeholder said he had fifteen minutes for me, so I set about establishing rapport as I did in all sales calls. Noticing a statue of a golfer on his desk, I launched into a conversation about the sport. He seemed to enjoy himself, but once the fifteen minutes were up, he looked at his watch and announced, "Oops, got to go, been a pleasure meeting with you." As I left, he invited me to set up another meeting. The problem is, he never again found time for me. Even though I was a "nice guy" and had great rapport, he probably wrote me off as someone who had wasted his time.

Small talk has its place but in my experience, CXOs want proof quickly that you can earn a spot in their busy schedules. If you don't provide them with such proof directly, you'll lose ground fast.

When you first meet with a CXO, shake hands, exchange pleasantries, and then follow his or her lead as to what should happen next. Do not launch into monologues about the weather or recent sports events. When you're given the opportunity, move straight into your objective: to convince your CXO that you can create real value for the company.

Financial Stakeholders tend to have rather large egos, so tread lightly. For instance, "Did you receive my e-mail?" is a bad opener since this stakeholder likely gets so many e-mails that he either did not read it or just does not remember it. Either way, the question implies he slipped up. Worse, your words may also imply that you believe there might be something wrong with his e-mail system, and by extension himself, for allowing the problem to exist. They may take this observation personally.

Because Financial Stakeholders tend to be visionary, they are usually not as detail-oriented as others. They have a staff to handle details. But their reliance on staff is good for you because if they like you, they are likely to hand you off to the appropriate person to evaluate the technical aspects of your solution.

Champion Stakeholder

The *Champion Stakeholder* is the person or people in or outside the account who want your sale to succeed. A Champion can help you keep the sale on track by introducing you to other stakeholders, evaluating your solution, or referring you to other buying organizations. An important point: Champions are made, not found. The seller must consciously look for Champions both before entering the account and while servicing the account, while the other stakeholders are already inside the account.

Champions are your internal cheerleaders, advocates who may even put their jobs on the line to further your cause. Having a Champion is much more than knowing that your brother-in-law or a friend of a friend works somewhere in the company. Your relationship with the Champion is professional and provides clear reasons for why both can benefit. In fact, there *must* be this reciprocity for the relationship with the Champion to grow. Knowing this, the seller should look for ways to reciprocate by increasing the Champion's job security or political power within the organization, or finding other ways that the Champion might benefit from the relationship. Otherwise, when the going gets tough, and it likely will, the Champion will not get going.

Sharing some kind of social connection may work but it has weak underpinnings. In fact, in cases where the sale fell through due to a lack of support by the Champion Stakeholder, it is likely the win/win relationship was not spelled out in advance. The seller should have an initial meeting with the Champion Stakeholder in which they nail down, up front, what each expects from the deal. It may be something as simple as an expectation that you will compliment him to the boss or it may be that a successful deal will bring him or her a promotion;

in either case, make sure the stakeholder understands and agrees with it. The payback for the Champion Stakeholder is usually intensely political and he may just want to bathe in the glory of a successful, high-profile project. In practice, it will be more complicated and you will have to be sensitive to political nuances.

I worked with a Champion who was several rungs below the Chief Executive Officer (CEO), and I asked him to set up a meeting with the CEO. He pulled some favors and arranged the meeting. After the meeting, I sent the CEO an e-mail thanking him for taking the time to see me, and complimenting the Champion on how well he was handling my account. I then got a private e-mail from the Champion who was awed that suddenly the CEO seemed to know and like him.

Most people want to be Champions. Often Champions are staff managers who report to the Financial Stakeholder and you are one of their assigned projects. Or, they may approach you, wanting to help you win. Either way, you have to find out what they want and how you can create a win/win long-term relationship, as soon as possible.

Technical Evaluator Stakeholder

The *Technical Evaluator Stakeholder* reviews new technology and refers the winning suppliers to the Financial Stakeholder. Because sellers are trained on the technical aspects of the solutions they provide, this stakeholder is often the first point of contact for sellers. I have found, however, that entering at this level only lengthens the sales process. Evaluators are usually friendly and knowledgeable, and they will often stroke the ego of the salesperson by trading technical information. The resulting pleasant conversations, by and large, are often a draw on your time.

The Technical Evaluator Stakeholder is essentially a roadblock, put there for the benefit of the Financial Stakeholder. Evaluators run interference, since they have the ability to say no but no authority to say yes. Because of the onslaught of sellers who approach them and the limited room for potential solutions, Technical Evaluators often come across as negative—not necessarily on purpose but out of necessity and self preservation. That is not to say that they do not have a vital role in the buying process. Their ability to protect the Financial Stakeholder's time as well as their valuable technical acumen make them key to the evaluation and selection process.

You should communicate with the Technical Evaluator after being referred by the Financial Stakeholder. Rely on your technical training, plus the political clout that the referral from the CXO gives you. It is appropriate to talk about the bells and whistles embodied in your solution, exchanging technical information and educating him on your competitive advantages. Technical Evaluators are very comfortable in "feature" level discussions. Be aware, though, that presenting the broader benefits (what the solution can do for the company) are best saved for the Financial Stakeholder.

Cost Evaluator Stakeholder

Cost Evaluator Stakeholders are typically the purchasing agents for a large account or the contracting officer for a government sale. Their job is to evaluate the price of the technology among the vendors and recommend a "vendor of choice." Normally, they look for three quotes, put you into price competition with other vendors, then recommend the lowest price to the Technical Evaluator or Financial Stakeholder.

Of course, if you are involved in a fairly new area of technology, there may be no direct competitor. In such cases, this stakeholder may be no barrier at all, as the Technical Evaluators will provide written justification for using your solution. Your challenge is to be the unique sole source, without competitive bids, and walk it through the purchasing process.

Cost Evaluators are generally not technically trained. Most do not understand what they are buying or why they are buying it. Because of this, the solution's deeper value to the organization is ignored. This makes the Cost Evaluator a potentially dangerous player, but bringing him into the sales process early—and keeping him informed—is a wise political move. At the appropriate time, he will be used to procure your solutions under the Financial Stakeholder's approval and direction.

End User Stakeholder

End User Stakeholders are those who actually use the solutions that you provide. They are very concerned about how their work performance will be affected by your solution, which make them influential characters in the buying process.

Your solutions probably affect too many people for you to influence these stakeholders individually. Sellers should depend on their firm's marketing efforts to create demand and educate the market at the end-user level. The users are, after all, part of the general audience that marketing is intended to reach. Trade shows, conferences, and networking seminars are especially effective at influencing users. If there is a large group of users within the company, there may be an internal special interest users' group that you can identify and work with.

If the End User Stakeholder is not sold on your product, buyers usually will not buy. And if they do buy without the end users' approval, the implementation of the solution will usually fail. Ignoring the end-user community can be a drastic mistake, one that will ensure there is no repeat business.

Putting the Major Account Profile Together

A few years ago, when I was just starting out in the field of studying the effectiveness of sales forces, I had an opportunity to bid on a sales training contract with a large manufacturer of personal computer products. The company had hundreds of salespeople all over the world, and senior management was looking to implement the I.T. Sales Boot Camp. My primary contact was the senior vice president of sales who, in this case, was the Financial Stakeholder in more ways than one. First, he had the budget to make the deal go forward, and second, the training directly affected him and the revenues of his field force.

The vice president told me there were two other competitors. Our strategy was to partner with the other two competitors and offer a one-stop solution to the company, thinking that with only one supplier bidding for the business we could not miss. There was plenty for everyone.

It was all too rosy. During the proposal writing stage, I was introduced to a "new" training manager. In short order, it was obvious by her actions that she viewed anyone from the outside as a threat to her turf. She spent her time trying to convince senior sales management that her staff could do this "in-house" and that consultants were an "unnecessary expenditure."

While the vice president liked our proposal and respected our approach, he backed off on giving us the go-ahead. The vice president of sales viewed this new manager as a long-term player and was not willing to overturn her decision to build the program and implement it in-house. So a sure deal was put on the back burner while the new manager went to work. To this day, many years later, the sales force is still floundering, and both the vice president and the manager are no longer with the company.

Without a doubt, we were "in" with the Financial Stakeholder but still were not able to move the business to closure. We did not have the support of all of the players—in this case the Technical Evaluator Stakeholder—and it sank the deal. In hindsight, we should have asked for the VP of sales to become the Champion Stakeholder and "sell" the Evaluator for us early on. We should have armed him with the proper tools and trained him on what to say and how to say it.

To help navigate through the political whitewater, consider filling out a Stakeholder Assessment Worksheet (see **Figure 1-5**) found at the end of this chapter. This tool helps you assess the strength of your contact with the stakeholders and identify those who are allies or enemies to you and your company.

Summary

Mapping your way through dangerous terrain is a daunting prospect to most salespeople. In fact, most avoid it, wasting their time on cold-calling without a larger plan and on other forms of undirected wheel-spinning.

By now you understand that it does not have to be that way, and that intelligent planning can start you down the path to productivity. But your planning must take your team into account as well. So who are the right players to assemble on your team? What should your team look like? How do you orchestrate a coordinated assault? This is the subject of the next chapter.

Resource File: Major Account Profile Worksheet

The purpose of the MAP worksheet is to organize strategic information gathered by the IT selling team. This information about the seller's solution is taken directly from the perspective of key stakeholders. Research conducted through some of the sites listed in the Sources of Information Appendix can be used to get some of the information. You can also glean valuable information from the Champion Stakeholder. Reviewing and revising the information on this worksheet should be an ongoing effort as the relationship with stakeholders deepens.

Figure 1-5: Resource File Stakeholder Assessment Worksheet

Stakeholder Assessment

Rate each question on a scale 1–10.

Stakeholder Ranking—Financial (F), Champion (C), Technical Evaluator (TE), Cost Evaluator (CE), End-User (EU)	Key Stakeholders Fill in the names and rank, one column per stakeholder						
Stakeholder Ranking	TE	F	C	CE			
1 2 3 4 5 6 7 8 9 10 Not at all Sometimes All the time	Bill	John	Mary	Joe			
1. I discuss topics that are not work-related with this stakeholder.	9	5	10	5			
2. This stakeholder recognizes the value that my solution brings to the buying organization.	8	7	10	5			
3. This stakeholder introduces me to others in the buying organization that have an interest in my solution.	10	7	10	6			
4. This stakeholder has business strategy for implementing my solution.	8	5	7	3			
5. I am the supplier of choice with this stakeholder.	9	5	9	4			
6. This stakeholder discusses key internal business strategy and inside information with me.	10	4	10	3			
7. This stakeholder relies on me for key business and technology information and treats me like a peer.	8	6	9	3			
8. This stakeholder can articulate my company's long-term strategy for building a relationship with his or her company.	7	7	8	2			
9. This stakeholder contacts me.	8	2	8	1			
10. I feel that I am developing a personal friendship with this stakeholder.	7	5	9	3			
75 to 100 = Buying Ally (BA) 50 to 75 = Advocate (A) 25 to 50 = Neutral (N) 0 to 25 = Enemy (E) **TOTALS**	84	53	90	35			

Chapter Two
Assembling the Pursuit Team: How to Create and Deploy

THEY HAVE SHOWED YOU your new desk, the location of the restroom, the coat rack, the photocopier, and the coffee machine. They've set up your e-mail address and your telephone extension, and shown you how to log onto your computer. You've shaken hands all around.

Have a seat. Adjust the chair. Now, put everything out of your mind except the item in the office with the crudest technology: the calendar on the wall. Because you have just a few months to make it, to sell and exceed your assigned quota.

The good news is that you have a team within your company who will help you succeed. The bad news is that you are expected to know how to deploy this team and make them effective from the moment you adjust your new chair. In a large, established company, many of the required team resources will be handed to you before you get through settling in. In a small firm, you may have to go out and lay claim to them. In a start-up, you may have to pull them out of the air. Whatever the situation, sales teams are critical to the success of IT sellers, so let's take a look at how teams are created and deployed.

The Power of Team

Companies such as Sun Microsystems, Xerox, and Oracle have pioneered approaches to organizing sales forces. These world-class sales organizations, using the I.T. Sales Boot Camp methodology, have adopted team-selling to manage sales efforts in large accounts. No matter how large your organization, you can use it, too.

A great example is found with Xerox, where team-selling on major accounts is a well-established system for integrating the company's powerful resources into the sales effort. For years Xerox built its selling teams around two basic ideas. First, that all members of the sales team must pursue common objectives. Second, that top Xerox executives are an integral part of the team.

AT&T is one of Xerox's largest customers, and its Xerox account management team numbers more than two hundred salespeople, most at the district sales level. Xerox has sixty districts in the United States for AT&T alone, and even more abroad. These team members comprise the ground troops in the sales effort. In addition, "business line" or "subject matter" experts from other divisions also report to the team. These include finance, administration, and technical specialists for specific products, services, or business areas. When setting up the team, members agreed on Xerox's sales objectives in four major areas:

1. Buyer satisfaction
2. Sales and profits
3. Satisfaction produced by employees
4. Best use of sales tools and coordination processes

Xerox is serious about this approach. The performance of each team member is based on these measurements, which determine their annual bonus compensation and advancement to other positions. Xerox also emphasizes common objectives in less formal ways, such as by asking team members to rate each other's performance in a questionnaire and to make specific suggestions for improvement.

It's not easy maintaining a productive level of communication among team members, ensuring that two hundred people who live and work in different cities and countries sell and work as if they have coffee together every day. But buyers are happy, and that's what counts.

Scratching the Niche

Traditionally, sellers are assigned geographic "territories" to cultivate new business, and most use a "shoot at anything that moves" approach in managing their patch of ground. But territory management is changing; in fact, there is a change in the very definition of territories. With the advent of the Internet and complex channels of distribution, this traditional notion of territories raises several problems. Who gets paid for the sale? How much are they paid? Who has influence with the account?

An "industry" or "industry/product matrix" approach is replacing the geographic division of sales territories. At the same time, technology sellers are dropping the old geographic divisions of sales responsibilities, shifting to "industry niche" and "product-line niche" to segment their markets. Instead of being assigned all of Toledo, for example, top sales-people are covering a particular vertical industry, such as finance, energy, or healthcare, which requires "business line" or "subject matter" experts. This new way of organizing sales teams brings about a shift to national marketing and team-selling for large accounts. Communications technology—such as Web conferencing or e-mail—makes a broader span of control feasible. With a few keystrokes, groups of people can be contacted or connected, passing information at lightning speed, thus making decision-making for buyers and sellers faster and more effective. This is truly a customer-centric sales universe.

A popular strategy is to combine the industry approach with geo-graphic territory assignments. This hybrid blends the sophistication and commitment of team-selling with the direct contacts that geographic territories make possible. Top accounts get a national team who visits less often, but they can still maintain constant contact via the best electronic tools.

Sales calls to large clients are, in effect, conferences among senior managers of both the buyer and the seller, everyone who is involved in using, making, selling, supporting, and marketing the product. They are followed up and reinforced by more frequent contacts at the telesales or inside sales level. These are the traditional calls to check on status, confirm buyer satisfaction, or fix any problems that occur. They also give the vendor a close look at how its products are being used or marketed by the buyer. In some cases, the "touching" function is basically a buyer service job. This is not a senior selling position and, in many companies, has not been considered a part of selling at all but of product service.

At IBM, a team of outside senior salespeople follows the industry or industry/product model. Buyer service specialists are assigned geographically, with great emphasis on helping them frequently contact as many buyers as possible. The critical skills of such junior service people involve attention to account management and a focus on the buyer's values. This information is then given to the outside account managers for face-to-face meetings with the buyer.

The tricky part of managing hybrid models of national and local selling is to work within a team where the lines of responsibility run in several directions at once: up, down, and crosswise. The territory can become a maze that no computer can navigate. Only resourceful and flexible sales professionals can figure out this new territorial geography and make it work effectively.

Organizing Your Team

Ultimately, any sales team is organized to successfully win deals. Each of the team members has a pivotal role to play in the sales approach and each individual's strength compensates for another's weakness.

Because of the complexity and the technical nature of the sales process, you must enlist internal experts with specialized functions. Some sellers believe they can handle these large accounts themselves. Swallow your pride: This type of selling requires continual resource coordination (see **Figure 2-1**). We all have weaknesses that we need to address, so instead of pretending to know everything, you should turn to your team composed of:

- Subject Matter Expert
- Executive Management
- Technical Expert
- Proposal Leader
- Graphic Designer

Figure 2-1: The Pursuit Team Wheel

Subject Matter Expert

The *Subject Matter Expert* has experience in the buyer's business and industry. With that knowledge, he is able to apply your company's solution or product to the buyer's business needs. He should be able to understand the requirements the buyer has presented and deliver the details of a solution that meet those requirements, in a lingo that the buyer will understand.

Finding such experts is not as hard as you may think. At some point, corporate executives retire and many become consultants, sometimes offering twenty to thirty years of expertise as industry insiders. You and your organization can use them. For example, if you have been assigned to the telecommunication market sector, you can contact the local Telco provider, speak to the human resources department, and simply ask for the names of executives who have recently retired or been downsized. The people in personnel will not think twice—this happens all the time.

Once you have a consultant lined up to be your Subject Matter Expert, he can help you in several ways. This expert can work with you in the MAP process, helping to plan which persons or divisions to target and how to go to market. They can bring you up to speed concerning the "who, what, when, where, and why" that is needed to understand the industry and its participants in advance of the initial contact. The consultant can provide introductions, not only within his old company but also within his former competitors. He can also accompany you on selected sales calls to provide the credibility that is necessary to gain trust with the Financial Stakeholder and Evaluators.

Ideally, the Subject Matter Expert will also have the insight to know which technical nuts and bolts are important to include in the proposal and which are not. Too much detail and the readers get bogged down; too little, and your solution is seen as superficial. Many Subject Matter Experts have a tendency to give too much detail. In order to be effective, they need very specific guidance from the sales team. Use them to create appropriate targeted responses to the buyer and to "differentiate" you from the competition.

Executive Staff

Sales and top executive management—*Executive Staff*—are also integral parts of your team. This includes legal, pricing, product, and senior management staff that the seller must coordinate to make deals happen inside the selling company. Including executive management on your team is critical. Without them, you will not have the resources to hire a subject matter expert, and you will have trouble securing the services of the other team members. These are the people who will fight for you when you sense that a product needs to be customized to make a deal fly, or when you see that the pricing or terms need to be adjusted to make the contract work for the buyer. You also need the top management on your team when it comes time to play good guy/bad guy in negotiating the contract particulars. The salesperson should never play the "bad guy" in the sales process; management must fill that role.

Technical Expert

You must try to become an expert on your product—there's no getting around that. You should be competent at giving authoritative demonstrations, and it is the *Technical Expert* who will bring you up to speed quickly. Your Technical Expert will often be a product manager, a founder, or another sales team member. In larger organizations, you may even be assigned your own Sales Engineer (SE). Remember that becoming an expert within your limited time to market is impossible, since even the simplest technology affects too many other products or services for any one person to understand. When a buyer asks about compatibility with some obscure piece of middleware, say you will get back to him with the answer, and then consult your technical expert. The buyer won't be put off; they probably have a list of questions and did not expect an immediate answer.

Proposal Leader

Meet the linchpin on your team. The *Proposal Leader*, known in some organizations as the project or program manager, is the one who knows the prospective buyer's hot buttons and the values it holds dear. Because he or she is the one with such knowledge, the Proposal Leader is really the only one on the team who has the overall big picture, the vision of how all the pieces of the proposal should fit together to carry the winning message. He is also the one who will ensure that hard numbers, and not new product features, are used to make the business case that appeals to the Financial Stakeholder.

The Proposal Leader must be sufficiently detail-oriented to follow complicated proposal formats dictated by the buyer. He must be a good delegator, able to dole out assignments that support the big-picture objectives. And he must show a knack for motivating people to keep the effort going when fatigue and weariness sets in.

Graphic Designer

The *Graphic Designer* is on board to hear what you want to say in the proposal or the presentation and then to prepare visual concepts that help deliver the message. Graphic Designers must be good listeners and have an open mind.

Selling Inside Your Company

While at Novell, one of my assignments was to manage a distribution sales team that was responsible for Tech Data, the world's largest PC networking distributor. The account manager I assigned to them was

Sharon, a real go-getter who did a great job developing rapport at all levels in the account. They loved her.

In fact, Sharon's magic reached all the way to the Novell corporate office where she spent lots of time and energy waving the Tech Data flag to senior executives. Sharon, who was a Novell employee, fought for Tech Data as if she was an integral part of the Tech Data team—and she was. Her endearing service caused Tech Data to be tied very closely to high executive levels within Novell, which spearheaded massive sales.

In the buyer's eyes, Sharon was indispensable. Time and time again, I see this same story with other experienced, best-practice sellers. These salespeople fight for both the buyer and the seller at the same time, making them indispensable.

As the "point guard" for the account, you will spend much of your time selling inside your own company. Some of your efforts will be directed at communicating your buyer's wishes and attitudes to those who do not have daily contact with the account. Product customization, pricing concerns, and sales forecasts are all challenges the seller faces each and every day in an effort to marshal internal resources for the buyer. This may not be the glamorous part of selling, but the effort is worth making.

As a part of sales activities, for example, salespeople are usually required to fill out forecasts for management about their accounts. Most hate doing this, as they feel mistrusted or view the reports as political and a waste of time. But in order to make sales teams work, sellers need to plan for the future. How many units to build or what revenue stream will be there to fund further development are just two of the many concerns when handling large revenue opportunities. Forecasts may be repetitive at times, but they provide the team with a guidepost on which to focus.

I was recently working with an Internet data storage company that showed dismal sales forecasts for its new service. Essentially, the forecast showed that the service was not getting any traction with large enterprises. By querying potential buyers, the company found that the large buyers already had the infrastructure that the new service was supposed to provide. They were not interested in the service, but they were interested in the software that powered it. To solve this potentially catastrophic problem, the company teamed up with one buyer to develop and install the software at the buyer's site, thereby establishing a new and sustainable revenue stream.

This story illustrates the importance of forecasting and planning and how it can be used as an extremely valuable tool for the buyer as well as the seller. It gives sellers an opportunity to look inside the pipeline and shift gears to keep the buyer happy.

SWOT Competition Analysis

Once you have your team in place as a group, complete a SWOT Competition Analysis. SWOT analysis is an examination of the buyer's internal strengths and weaknesses and external opportunities and threats. Such an exercise gives you the best shot at a grounded plan from the *buyer's* perspective.

Selling organizations can use the SWOT analysis very effectively. In fact, leading vendors like Agilent Technologies use the SWOT analysis to plan and manage relationships with potential and existing accounts. Sellers are asked to provide sales management and selling teams with analysis on a quarterly basis for all forecasted sales. This information is shared with all parts of the sales team to give a broader perspective of each opportunity.

Buyers can use this instrument as well. For example, an information technology department needs to determine the internal strengths and weaknesses of its people and its technology. It also needs to ensure the IT strategy complements the company's overall business goals. The department head needs to know: What strengths does a strong staff member bring? Where can he improve? At the same time, stakeholders must consider the organization's external opportunities and threats, from buyers and competitors. How attractive is the market or direction they're considering? What's their market share and cost structure?

Dell Computer is a great example of an IT company using a SWOT analysis to carve out a strong business strategy. Dell recognized that its strength was selling directly to consumers and keeping its costs lower than those of other hardware vendors. As for weaknesses, the company acknowledged that it lacked solid dealer relationships. Identifying opportunities was an easier task. Dell looked at the marketplace and saw that buyers increasingly valued convenience and one-stop shopping and that they knew what they wanted to purchase. Dell also saw the Internet as a powerful marketing tool. At the same time, Dell realized that competitors such as IBM and Compaq Computer had stronger brand names, which posed a threat and which put Dell in a weaker position with dealers.

Dell put together a business strategy that included mass customization and just-in-time manufacturing, in other words letting buyers design their own computers and then custom-building the systems. Dell also stuck with its direct sales plan and offered sales on the Internet.

Where does the IT selling pursuit team fit in? Here are some questions that the Proposal Leader can use when leading the sales team in a SWOT analysis of a buyer.

Strengths (Internal)
- What does the buyer's company do well?
- How strong is the company in the market?
- Does the company have a clear strategic direction?
- Does the company's culture produce a positive work environment?

Weaknesses (Internal)

- What can be improved in the company?
- What does the company do poorly?
- What should be avoided?
- Is the company unable to finance needed technology?
- Does the company have poor debt or cash flow?

Opportunities (External)

- What favorable circumstances does the company face?
- What are current trends? Is the company positioned to take on those trends?
- Is the company entering new markets?
- Is the company advanced in technology?

Threats (External)

- What obstacles does the company face?
- What is the competition doing?
- Are the required specifications for the company's products or services changing?
- Is changing technology threatening the company's position?
- What policies are local and federal lawmakers backing? Do they affect the company's industry?

SWOT - Account Analysis

Account Name:		Date:	
Account Manager:		Value:	
Systems Engineer:		Close:	

Stakeholder Type	Name	Title	Covered 1=Low 5-High	Primary Contact Person Responsible for the Buyer Contact
Financial Stakeholder*				
Technical Evaluator Stakeholder*				
Cost Evaluator Stakeholder				
User Stakeholder				
Champion				
Champion				

*Financial Stakeholder is ONE person who can say YES when all others say NO
*Technical Evalutor Stakeholder can say NO but cannot say yes

Customer Opportunity: What problems are they looking to fix, change, or avoid . . . and WHY?

Competition: SWOT - Strength, Weakness, Opportunity, Threat

Strength	
Weakness	
Opportunity	
Threat	

Competition: SWOT - Strength, Weakness, Opportunity, Threat

Strength	
Weakness	
Opportunity	
Threat	

Action Plan:	Date Complete	Expected Outcomes

Summary

By now I hope I've convinced you that no IT salesperson is an island. You should have a greater appreciation of the suite of skills and personalities needed to put together a winning sales team, and the ways in which this ensemble can be deployed both inside and outside your company to achieve success. You should focus on spending time as a team, working on forecasts and SWOT analyses. Now that you have your team in place, it's time to take a closer look at the personalities of the people sitting on the other side of the desk—the buyers.

Resource File: Gauging Your Team Readiness

Do you have what it takes to sell as a team? To find out, answer these questions. When you are finished, add up your yes and no answers.

Your Team:

1. Do you have an effective system for people on your team to work together?
2. Do you have effective means of communication even if team members never physically meet with each other?
3. Does your management encourage cooperation rather than competition?
4. Have you created an open atmosphere in which team members are encouraged to share information rather than guard secrets?
5. Is communication between upper management and all potential team members lively and positive?
6. Does your company have a team-oriented commission plan?

Your Company:

7. Do you have a salesperson assigned to each account?
8. Do you have people who deal with specific buyers but who do

not have direct sales responsibility for them?

9. Do you have people with special subject matter expertise who could be helping specific accounts?
10. Do your people know as much about your accounts' business as they do about your own?
11. Do you have access to people who could help you learn more about your accounts' business?

Motivating Your Team:

12. Have you established financial incentives that reward all team members?
13. Do you have a reliable instrument for measuring buyer satisfaction?
14. Do you have a way to link buyer satisfaction to team performance?
15. Do you develop cooperation within the team through effective leadership?
16. Do you encourage team members' creativity with buyer problems on site?

Total:_____ YES=_____ NO=_____

How to Interpret Your Score

If you answered yes to at least ten of the previous questions, you are ready to start using the advantages of a team approach right away. To strengthen your position, concentrate on the areas where you answered no.

Chapter Three
Preparing for the Attack: How to Get into the IT Buyer's Head

AS A SELLER, you have always relied on your personality to carry the day. But in your first appointment with a potential buyer, it is apparent that he does not like you. For no good reason he is not listening and is downright rude. As much as you try to change your approach, you're dead in the water.

The next day you walk into an appointment with a new prospective buyer and everything pulls together splendidly. It's as if you have known this person all your life. You and the buyer are like two peas in a pod. She laughs at your jokes and likes your style.

What dynamics are at play to explain such different reactions? Regardless of how technical your solutions are, you are still working to influence the sales decisions of other human beings. It is often said that buyers buy from sellers whom they like, or want to be like. This is especially true in a competitive marketplace, where the solution can be found from many sellers. Unfortunately, all buyers are not like you. Still, you must find techniques to create a selling environment in which the buyer is comfortable enough to make a decision to move ahead with your proposed solution.

Why Buyers Don't Buy

For years sellers have desperately sought answers to the following puzzles: "Why don't they call me back?" "How come they don't buy?" Let's take a look at Jim, an IT seller. Jim is a good salesperson—he speaks the language and understands the concepts as if he were born with them. He knows his company's capabilities and has been trained on the products and services they offer. He dresses well, smiles, and looks the buyer right in the eye. So why does he get such poor results?

This is what happens on a typical sales call with a prospective buyer, during which Jim makes a wonderful presentation. He talks about capabilities, company strategy, and technologies he has to offer. The buyer flips through the collateral materials Jim provides and asks him about his company's service. "We can do pretty much anything the buyer needs," says Jim, who goes on to say that they "customize" each solution to "meet the buyer's needs." He adds that he has three other large buyers who are very satisfied with his firm's service.

The buyer asks about how the sales process works. Jim tells him that his consultants are the best, and that they were recently written up in a local publication as a reliable and fast implementer.

The buyer then asks if he can handle software installation. Jim replies that installation is a breeze and that all of the software is guaranteed to work. "Any end user will appreciate this product," he says confidently.

The buyer puts the collateral materials on his desk, thanks Jim for his time, and walks Jim to the lobby. We last see Jim leaving the building muttering, "Some people are just difficult to sell."

No questions, no empathy, no listening, no understanding, and poor assumptions—everything in this sales call is wrong. Jim does not directly address the buyer's questions regarding the technology, nor does he pick up on the buyer's concerns about installation. Such a sales interaction is common with new and experienced sellers alike. The new ones probably don't know better, while the experienced ones know the script too well. Contrary to the assumption that underlies Jim's approach to the conversation, buyers do not buy because they are made to understand the solution. They buy when they feel that they have been understood. Successful salespeople show that they understand the buyer and give them what they want in an understanding way.

Understanding the Buyer

Buyers bring to the sales process personality styles known as "observable behavior." Their behavior determines how information is received, how decisions are made, and how comfortable they feel about doing business with you. Face it, you would rather do business with people like yourself, as would your buyers. It's more comfortable, more familiar. You would intuitively understand their thought processes, their motivations, their interests. Of course, you can't count on this happening in a given sales situation, so you have to empower buyers by creating an environment that will be comfortable for them, one in which their motivation to buy can be expressed. Doing this requires an understanding of the observable behaviors that both you and the buyer bring to a relationship, and how they work together.

Consider the basics of your own style. For instance, how do you approach an appointment with a prospect? The way you do it constitutes one of your observable behaviors. For instance:

- Are you prepared, concise, and businesslike?
- Do you approach the buyer in a warm, friendly manner, looking to build personal rapport with little concern for business details?
- Are you relaxed, patient, methodical, and prepared with details?
- Or have you come well prepared but feeling uneasy about building rapport?

With respect to the prospective buyer, what behaviors did you observe during your meeting?

- Is the prospect direct, impatient, and to the point?
- Does the buyer appear friendly, personally interested in you, and easy to approach?
- Is the buyer relaxed, patient, and attentive; task-oriented with strong team affiliation?
- Or is the buyer precise, detail-oriented, tense, and critical?

Such observable behaviors can tell you how to proceed in the sales call. As the stakeholder responds, observe his behavior in such a way that you can map your strategy to create effective communication, one that will cause him to want to buy. This strategy must take into account your own observable behaviors so that they do not clash with those of the buyer. Buyers are more likely to make an open emotional response when they are comfortable with the seller. Of course, they will be most comfortable with sellers whose observable behaviors are similar to their own, but there are ways to handle dissimilarities, if you are aware of them.

Be like a detective who is sensitive to behavioral clues left by the stakeholder, and be aware of the ones you are sending out as well. The ultimate objective is to create an environment in which stakeholders can recognize their own needs, realize that your solution is a match, and commit to moving to the next step in the partnering process.

Four Types of Buyer Behaviors

The foundation for using observable behavior is intelligent listening. The art of "active listening" is covered in many generic sales training books. IT sellers need to bolster what they know with specific observations about four behavioral types (see **Figure 3-1**) commonly found among IT stakeholders: the Take-Charge, the Optimist, the Amiable, and the Analytical personality. Each requires different handling. You must also be aware of your own behavioral type, to use it as an asset. Understanding how these behaviors interact, along with other tactics found in this book, will go a long way in ensuring sales success. The better you are at becoming a chameleon that adapts to different personality types, the more successful you can expect to become.

Figure 3-1: The Four Types of Buyers

Behavioral Traits

Take-Charge	Optimist
• demanding	• enthusiastic
• egocentric	• inspiring
• ambitious	• political
• competitive	• persuasive
• decisive	• polished
• forceful	• warm
• pioneering	• trusting
• aggressive	• sociable
Amiable	**Analytical**
• relaxed	• careful
• passive	• accurate
• steady	• diplomatic
• team-oriented	• worrisome
• possessive	• evasive
• deliberate	• rules-oriented
• stable	• tactful
• predictable	• neat

The Take-Charge Personality

Take-Charge types are action-oriented and decisive: born leaders. They are quick on their feet but sometimes socially awkward, which probably explains why so many Financial Stakeholders fit this behavioral pattern. In fact, their primary fear is of being taken advantage of by the seller, which means they are automatically wary and occasionally rude. You can handle their wariness by presenting yourself as a peer and a consultant, one with integrity who is well prepared and who gets to the point. It is good to know all you can about the company and its present position in the market. This feeds into their ego, and for Take-Chargers, ego is strongly associated with professional success.

Here is how you can identify buyers who exhibit the Take-Charge personality:

- They often make reference to being busy or to "time constraints." This is their way of letting you know that they are important.
- They walk fast and are often physically competitive.
- They interrupt you if they think they are being given too many details.
- They will openly show impatience with others.
- Their speech is direct and brief.
- They are very results-oriented and want assurance that the proposed solution will get them to the desired end efficiently.
- They look for shortcuts to the end result.
- They may break rules to get their desired results or make new rules.
- They are multitask-oriented.

When talking with these types, be straightforward, use facts, and focus on business. Dispense with small talk. Do not rely on documentation, details, or technical information. Never show up late; if you do, it shows a lack of respect, which they take personally.

Don't infringe on their ego by asking questions that could have been answered by other sources prior to your meeting. When it is obvious that you have done your homework prior to the meeting, the unconscious message that is sent is that you value their time. Some sellers send a brief and concise agenda for the meeting via e-mail a few days prior, asking the Take-Charge buyer to add to or subtract from it. This will show that you value his time and demonstrates organization, something he himself is very good with.

If you observe several of these characteristics, either during a telephone conversation or in person, tailor your communications to match the buyer's tonality, words, phrases, or gestures. This will increase your chances of influencing the commitment.

Take-Charge buyers are not afraid to make mistakes. This key attribute can lead to a shorter sales cycle, since they are less likely to be afraid to commit to a solution. Because they usually have a better understanding of their own needs than other personality types, they will quickly know whether or not your solution will help the organization.

The Optimist

Optimists or "Os" are people-oriented and tend to measure things according to relationships rather than numbers. They crave social interaction, hate rejection, and are less competitive than Take-Chargers.

"Os" are sometimes found at the Financial Stakeholder level, but are more typically found among Evaluator and Champion stakeholders. Optimists make great Champions because of their willingness to look at the bright side of technical and political changes. Typically, their people skills help them land comfortable staff jobs without the burden of authority. If the Financial Stakeholder is an "O," it is best to sell to him outside the office—at lunch or at the golf course. Don't forget the old IBM sales caveat: "The buyer must not be forced to drink alone."

Here are some characteristics of the Optimist's personality:

- They seem talkative and friendly in conversation.
- They often have poor listening skills and display poor retention of detail, yet can be very articulate.
- They enjoy social recognition and awards.
- They can be creative when solving problems, without being bound by traditions or others' opinions.
- They often seem enthusiastic and excitable.
- They motivate others by being persuasive rather than logical.
- They can be humorous and entertaining.
- They avoid facts and figures.

The Optimist values trust above all else. You should establish trust as quickly as you can and focus on keeping it. If the Optimistic buyer thinks you have broken trust with them, it will be exceedingly difficult to win it back.

Optimists show their emotions easily. Because of this, they are the easiest to influence—these are the folks who buy impulse items on display racks beside supermarket checkout lanes. It is often said, "Salespeople are the easiest sell." That's because sellers themselves are often Optimists and find other "Os" comfortable to deal with.

A pitfall for bringing two Optimists together is that they will have a "good meeting," during which they will accomplish little beyond creating mutual rapport. Your job is to keep the Optimist focused so he makes a defensible decision, not one based on emotion alone. It is said that there is nothing stronger than a person who believes in something and who is also correct in that belief. Make sure the Optimist is correct.

The Amiable

Amiable types are team-oriented at work and family-oriented at home. More than half the people you will encounter in the typical corporate organization will fit this behavior pattern. The fact that they are amiable is what keeps many of them out of upper management; I cannot recall encountering one amiable personality among CXOs, but I have seen a great many among the Evaluator, Champion, and End User stakeholders, probably because their personality style is steady, calm, and emotionless.

This is what else you should know about the Amiable personality:

- They are loyal and will stay with one employer for long periods.
- They maintain long stable relationships and serve their leader steadfastly.
- They are patient and always appear relaxed.
- Although they are emotional, you will not be able to observe it.
- They appreciate being part of a "team" where they can serve and make a contribution.
- They tend to resist change. They can be so attached to a situation that they become possessive and identify with it.
- They are indirect in speech and avoid confrontations.
- They are not inclined to take risks and they admire reliability.
- Their offices will have pictures of their family and teams.

To reiterate, Amiables are not comfortable with sudden change and tend to be very loyal. This can work against you as a seller as they may have loyalties to your competitor; but it can work for you if you can win them over.

Be prepared to give them lots of information, as they require substantial time to commit to a solution. They are patient, listen well, and need to know that the solution is reliable. They do not want to be responsible for a solution that may turn out to be "unreliable." Further, because it is often their responsibility to be Technical Evaluators, they do not want to recommend any solution that is not going to be beneficial to the organization as a whole. To champion your solution, it must be clear to them that they will not regret their decision in the long term.

Since they are team-oriented, you should use your team as much as possible in meetings or make references to them in conversation. Amiables are best swayed by written proposals that emphasize the quality and breadth of the organization that you have behind you as well as the organization's past achievements. Demonstrating past performance is important because these types are not risk-takers. In today's dynamic and changing IT world, you may well be representing a smaller company, making Amiables uneasy. Their immediate thought is that you cannot support your solution, a common objection that reflects their behavior style. To overcome this objection, be prepared with success stories and references.

Because they resist change, Amiables present a sales challenge. Ask them what they like about their present way of handling challenges and reassure them with similarities before emphasizing differences. When they have revealed the shortcomings of their present solution, then link their best conception of a solution to what you have to offer. Reassure them that your solution has already experienced great success. If you represent a small company, emphasize personalized service as a form of reliability.

An Amiable is not as trusting as an Optimist but is more people-oriented than a Take-Charger. You want him to know that you will earn his trust over time and that you are willing to put in that time. Winning over a loyal Amiable is an excellent investment of your sales efforts, since such persons almost always enjoy the trust of the CXO for technical recommendations.

The Analytical

Call them nerds, geeks, dweebs, or bean counters—just keep it to yourself. Analytical types are detail-oriented. They find out what the rules are early on and follow them. They are not particularly interested in making the rules themselves, only in meeting all parameters set by their organizations. They need more information than other behavioral types to make a buying decision or even to put their stamps of approval on a solution. Before they give approval, expect analyticals to be critical and ask detailed questions, often exhaustively. Given their technical acumen, they are typically Technical Evaluator or End User Stakeholders. I have never come across one at the Financial Stakeholder level, but if you do, you had better be meticulously prepared.

Here are some Analytical personality characteristics:

- They want to understand all required parameters.
- They prefer that others set those parameters.
- They speak factually, accurately, and precisely.
- They work methodically and systematically.
- They approach tasks with high standards and expectations.
- They seem most critical of themselves.
- They are restrained and diplomatic in relationships.
- They fear making mistakes and look for factual reassurances of success.
- They appear unemotional during interactions.

In the IT world, many engineers fit this profile. It is important to recognize there are more Analyticals in information technology than in many other industries. Sellers must respect their expertise: Analyticals effectively link solutions with organizational challenges. This requires extensive industry and product knowledge. If you are inexperienced and cannot speak with this buyer at the necessary level of expertise, defer to one of your team members.

Analyticals represent quite a challenge for most salespeople, who interpret their natural behavior as "unfriendly." Do not take this behavior personally and don't avoid interaction with them. Instead, hold their hand through the sales process. Ask probing questions at length to uncover priorities, as it will take longer for them to reveal what motivates them. If your solution meets the established criteria, you will most likely win their support; but if you are missing some element they feel is necessary, you should emphasize the overall strengths of the solution and its application.

You can sway an Analytical with documentation—massive documentation, arranged in neat order. Count on them to read every word to completely understand all aspects of the solution. Additionally, keep in mind that the more analytical the stakeholder, the more they are afraid of commitment and change. To counter this, your documentation must stress that your product embodies the accepted "best practices" in the industry, and be prepared to prove it with testimonials from happy users willing to be references. If there is a herd to join, Analyticals want to know.

While many stakeholders ask for such references, the Analytical personality will actually follow-up and contact them. I once had an Analytical ask me for five references. He contacted four who gave very positive reviews of the success experienced as a result of our partnership. The fifth had moved on to another company that very week and the past company would not reveal the new employer. My prospect then contacted me to ask for another name as a fifth reference!

Mixing the Buying Types

So far you've got a good sense of where you and your buyer fit in the four personality styles. As you get to know these styles it becomes clear that you are not just one of them but a mixture of all of them. You probably see traits in all four styles that apply to you or your buyers. In all cases, one of the styles is the dominant "core" style for a person while the other three styles take a "subcore" role.

We are a combination of styles and these combinations can be pin-pointed to reflect how you or your buyer will act in a core selling situation. These core situations are the emotional charged times when the tough questions are asked in the sale or the uncomfortable decisions are made. It is in these times that the core style surfaces. Alternately, think of these styles as an artist chooses a color. There are many shades of colors in which to choose. Red, for example, could be pink, burgundy, or wine. Yet we think of all of these shades as being red.

As such, we are a blend of all four personality styles, one being the dominant style and the others are the shades. To find out where you fit you should mix the styles. Each of the four styles contains a mix of three additional styles as shown in the figure below.

Figure 3-2: Mixing the Four Types of Buyers

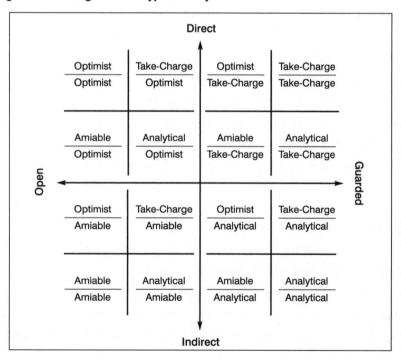

The Seller as the "Take-Charger"

If you are a Take-Charge type of person, you are characteristically impatient, impulsive, and to the point. Many salespeople fit this mold. You have a tendency to lose patience with an Amiable or an Analytical buyer because he wants and needs more information and documentation than you would use to make a decision. You also may push for answers with an Optimist without creating enough rapport to make him comfortable with you. That lack of rapport may strain the atmosphere, causing the buyer to feel that you are pressuring him and trying to "sell" him. If you are fortunate to meet a buyer who is also a Take-Charger, you will experience the bliss of instant understanding.

Recently, I met with a Take-Charge character in a potential buyer who was obviously avoiding me. Being a Take-Charger myself, I made a telephone appointment via e-mail. The appointment was for the following week, the person having indicated that he would have some free office time that afternoon. Recognizing our similar tendencies, I sent an agenda prior to the call.

During the telephone conversation, I asked several pointed questions, arranged to get the information I needed, thanked him, and hung up. The entire conversation took less than ten minutes and all other communication required only e-mail. Later, when we met face-to-face, he revealed rather sheepishly how pleased he was with our effective conversation. He hated to waste time and had expected me to take too much of his. What he did not realize—but that I did—was that we were on the same behavioral track—and that was my, and ultimately his, advantage.

The Seller as the "Optimist"

If you are an Optimist, or even if you have some optimist tendencies, you may think that building rapport is central to the whole sales process. Often this is true. But if a Take-Charger offers you a fifteen-minute first meeting and you spend ten minutes of it creating rapport, there will be little time left to "create value" for the stakeholder. It will then be very unlikely that you will get another chance to do so in the near future. With a Take-Charger, creating value is critical, so build value before you try to satisfy your desire to build rapport.

If you are an Optimist and are meeting with another Optimist, remember that you must stay in control of the conversation and keep the buyer focused on answering questions, clarifying needs, and understanding how your solution interfaces with them. Create rapport, but stay on track, as both of you may tend toward friendly, enthusiastic conversation.

With Amiable and Analytical buyers, your tendency will be to think that they are asking for too much information and are nitpicking or are even critical. On the contrary, this is simply what they need to make a commitment. Go with the flow by being prepared with documentation, even if you think it is overkill. Be careful not to give the impression that your friendliness covers up a lack of product knowledge. One technique is to ask yourself during the call, "How can I create value for this buyer?" This may keep you on track to uncover the important facts of the call, and keep you focused on the buyer's wants.

The Seller as the "Amiable"

If you are an Amiable sort, your danger with Take-Chargers is that you will give them too much information and they may become impatient. You will have pride in your solution and want to present all aspects. Instead, focus on just those areas that are beneficial to this stakeholder. There is plenty of time in the future to present your solution's other capabilities.

When the buyer is an Optimist, he may be uncomfortable with you because you have a tendency to not show your emotions as readily as he does. You are emotional, but you keep it in check. When selling to the Optimist, use your team instincts and "show" the buyer what he is buying. Keep in mind that the Optimist is an entertainer and loves a good show.

You will relate to Analyticals because you will understand their need for information and their reluctance to commit for fear of making a mistake. Your patience will be appreciated and will create confidence.

The Seller as the "Analytical"

Analyticals are often not in sales because it is such an inexact "science." If you have an Analytical personality, it is likely you have an affinity for detail and an interest in how things work together from a technical perspective. Sellers with this style tend to have an extensive product knowledge and will drive to give too much information, thereby boring the buyers or presenting so much that they cannot absorb all of it. Remember that the stakeholders are only interested in those areas of your solution that relate directly to their wants and needs. The Take-Charger will become annoyed and impatient with information overload, while the Optimist will stop listening and become bored with what he perceives as too much information.

Be careful with Amiables and your fellow Analyticals. You may get caught up in a lengthy information exchange without accomplishing any steps toward a commitment to buy. Analytical sellers get caught in the enthusiasm over the many fascinating features of the solution. This may lengthen the sales process by providing more information than is relevant, causing the buyer to look at new and previously unrevealed risks. The technical aspects of your solution are no doubt fascinating, but you must link them to the buyer's specific wants. Stay focused on the relevant facts.

Summary

In the world of technology sales, information spreads at lightning speed. In the middle of this avalanche of information, sellers should focus on the buyer's human characteristics, motivating the buyer in the ways that they want to be motivated.

Most technical people downplay the need for such "soft" skills. All the technology in the world, however, still has to go through human channels at one point or another. Integrating this information and applying it to your selling process will increase your value in the eyes of the buyer and have a dramatic effect on your win rate.

Next up: following the trail to the Financial Stakeholder.

Chapter Four

The Hunt for Money: Setting Your Sights on the Financial Stakeholder

"I HATE PROSPECTING!" Many times over the years I've heard that same complaint from well-meaning salespeople. The complaint is usually followed by a variation on one of the following excuses to justify their sagging sales performance: "I don't have the time," "I don't have good leads," "We are in a bad market." Are these reasons legitimate? I believe many sellers are just not cut out for the rejection that cold-calling can bring, that they don't possess the required discipline. As a result of being rejected so much, they find excuses for not wanting to continue. But the real question is: Why are they being rejected?

There are two reasons that make sense. The seller is calling on the wrong people, wasting valuable time educating the Technical Evaluator on the technical merits of the solution. Or having reached the Financial Stakeholder, the seller is intimidated by his customer's business savvy and doesn't know what to say and how to relate.

With IT sales, it is critical to enter the account at the Financial Stakeholder level. In this chapter I'll explain how to go about that and what to do when you get there.

The Rules of C"X"O Engagement

When I say, "Enter the account at the Financial Stakeholder level," what I mean is to have a face-to-face meeting with the specific CXO or Line Manager (someone who manages people and not projects) who is a Financial Stakeholder and who can make economic decisions to purchase your solution.

There are many ways to identify the Financial Stakeholder of an account. Of all the ways to try to get a meeting with a CXO, "cold-calling" is probably the most often used but the least effective. I know, at some level, you are delighted to hear this, since most sellers hate cold-calling. Most of the time you are turned away and are lucky to end up meeting with a Technical Evaluator, educating him so that he can look good to the Financial Stakeholder. Sending an introductory letter or e-mail prior to a call is not likely to improve the odds of getting a meeting with the CXO. And, contrary to popular belief, most CXOs prefer not to be introduced to sellers through referrals from outside their organization. Typically, the suggestion to meet with the salesperson is best received when originated from within the buyer's organization.

The most common reasons that a Financial Stakeholder takes a call or a meeting are, in order of effectiveness:

Existing relationship. If you have already sold something to that person, he knows your work and your reputation and trusts you. For these reasons and more, he will be glad to meet with you again.

Internal referral. Champions are an excellent way to find new decision-makers. They thrive on seeing you win and are a great resource. Also, an existing buyer may have no more business for you today, but perhaps he knows a Line Manager in another department with similar needs.

Company reputation. Some companies' reputations precede them, which is a significant competitive advantage. If you are from an industry leader such as Oracle, Microsoft, or IBM, almost anyone will take your call. There are very few firms in today's market with consistently reliable name recognition, but if your firm is the leader in a niche, use its name up front. On the other hand, if you are from an unknown or start-up firm, it is best to confidently enunciate your firm's initials, as if the buyer should know them. Using initials creates a neutral atmosphere and avoids the awkwardness of the other person not recognizing or, worse yet, prejudging the name of your firm.

Solution reputation. Market leaders sell reputation and you can too. If the solution you sell occupies a hot technology space, such as the business intelligence or Internet security marketplace, or if it offers significant value, such as the latest feature "hot buttons," mention the attribute up front instead of the company name: "I represent a company that provides . . ." If the account is in the market for this functionality, it may take your call.

The Power of Networking

The power behind word-of-mouth marketing has never been stronger. Buyers are relying on personal referrals and testimonials of people they know to help them make buying decisions. Studies in the IT industry show that up to 85 percent of all buying decisions are based on recommendations from others, such as business associates. Fewer buyers, meanwhile, depend on editorial and advertising sources for decisions. What this illustrates is that people rely on word-of-mouth information because it is based on their existing relationships and the credibility of the person passing the word along. It increases their sense of comfort with the decisions they make. The simple fact is that a positive mention of your name brings an increased level of confidence and convenience to your prospects, giving you greater access to them.

Of course, all word-of-mouth referrals need to be positive. Any negative word-of-mouth information will do considerable damage to your marketing efforts. Regardless of how you may perceive yourself as an advisor, it's the perception of the *buyer* that ultimately determines your success within the market. From my own experience over the years, I have discovered that three referrals typically result in one nearly immediate sale. The more referrals you get, the easier it is to build the high-quality relationships necessary to earn the right to present your solution. Here are several tactics you can use to acquire referrals effectively.

Practice Referral Talk

Have a well-honed referral gathering presentation for use with every potential buyer. This will increase your comfort, confidence, and competence. It also makes it easy for buyers to understand your needs and motives. For example, you could say, "If you feel that I've been helpful and professional, I would like you to tell people that you are my client. This will allow me to help those people just as I helped you." Practice your referral talk with your family, friends, and colleagues to sharpen your script so that it sounds natural and comes easily.

Let Them Know You Need Them

Recognize that everyone benefits from referrals. It's very important for your company's stakeholders to realize that your goal is to reach and work with more people like them. It makes it easier for them to refer you to people they know. Tell them that you want to continue to specialize, that you enjoy working with them, and that their personal referrals will give you an opportunity to serve other people in their market. This lets them know that your motives are to serve and sell other individuals in your niche in the same professional way that you've serviced them. It will also make it easier for you to ask because you are focused more on them and less on yourself.

Spoon-Feed the Contacts

It is not always easy for someone to think up some names of people they know in your niche without help. I found that it's much easier to prompt and stimulate the individual to give you referrals by giving them the names of four or five people in the market that you are going to be calling on. Use a membership list or directory of the people in your targeted niche market. Show them the list and say, "These are the individuals that I will be calling on. Which ones know you on a first-name basis?" Usually, they will select a name from the list. Then ask, "Whom have I left off this list?" This way you will acquire additional names right on the spot. By using membership lists from your niche market clubs, associations, special interests groups, and other targeted networks, you are sure to tunnel deeper into your market.

Prompt Them to Verbalize It

It is usually difficult for buyers to think of a referral because they have never really thought about it. You may want to ask them, "How do you feel my solutions have helped you?" Ask open-ended questions related to your products and services. Be sure to ask in a way that stimulates stakeholders to think about what value you have brought to them, and *then* ask for their referrals.

Ask Early and Ask Often

Sometimes prospects will tell you they are not ready to buy now. When that happens, ask for referrals even though they have not yet bought from you. They simply need to feel confident about you and your solution. Let them know that the reason you're asking is to be able to build a high-quality relationship with more people like them.

There is a lot of confusion as to when and how often you should ask. In reality, you should ask each individual several times. By asking more than once, you create a cumulative effect and will get greater results. So ask at the first appointment, at the time of the sale, during the evaluation, at the time of installation—all the time. Integrate asking into your modus operandi. Make it consistent and as natural as breathing. If it is unrehearsed, appearing unnatural, it may come across as pushy or, worse yet, desperate. The key is to ask early and set the stage.

Another great time to ask for a referral is when you send a thank you note for the business. Of course, if you've already gained referrals during the sales process, you would not normally ask again immediately in the thank you letter, but it doesn't hurt and can show your desire to serve others like them.

If networking does not work for you, then prepare for cold-calling. And then a new challenge arises: dealing with the gatekeeper.

Getting Through the Gate

Telephone cold-calling begins with getting through to the CXO on the telephone. But just calling the corporate switchboard and asking for him rarely does the job, since whoever answers the phone is likely to be the CXO's "gatekeeper." That person—whether you call him or her a secretary, receptionist, executive assistant, or even a gatekeeper—has the ability to keep you from talking to the right person.

Gatekeepers present many obstacles to your success because they are trained by the CXO to run interference and screen calls. Consequently, they are very practiced at saying "no." And they will say "no" if you give them any reason to do so or if you just seem unimportant enough that there will be no repercussions if they turn you away. Therefore, your objective is to impress upon the gatekeeper the importance of your call while giving no reason to say "no." The way you avoid that is by making sure you offer the absolute minimum amount of information required to talk to the stakeholder.

The main consideration is to avoid trying to sell your solution to the gatekeeper as a way of convincing him to let you through. You are wasting your time, since the gatekeeper is not the Financial Stakeholder. And you are putting yourself in a vulnerable position, since the gatekeeper can still say "no" to any and all of your arguments and prevent you from ever getting through.

Gatekeepers are trained to ask a lot of questions, to "qualify" the quality of the call. Consequently, they feel they are doing a good job only when they ask you a lot of questions, trivial though they may be.

But it is only after the gatekeeper is satisfied of having done a good job that you stand a chance of getting in—and then only if you have not offered any information that gives him an excuse to turn you away. Since the gatekeeper will be asking questions to decide whether or not to let you talk to the boss, the result is like a verbal game of tennis, with you serving and the gatekeeper returning with a question.

Suddenly, asking questions to control the conversation works in reverse. Normally you want to ask questions to elicit information from the prospect. Here, the gatekeeper wants an unspecified amount of information from you, and you want the gatekeeper to do one specific thing: put you through to the CXO.

When talking with the gatekeeper, the safest course is to respond to each question that you are asked, but then follow up with a directive. Then you use a question to confirm that the gatekeeper understands your directive.

And never lie. If you lie to a gatekeeper or to a Financial Stakeholder, you lose all credibility. I know that some salespeople have told tall tales to get through to a prospect. You are much better off sticking to the straight and narrow. But provide only the minimum amount of information to do the job. The less information you give, the fewer reasons the gatekeeper has for saying "no" and the better your chances of getting put through to the prospect.

Keeping all this in mind, here is an example of a conversation between a seller and a gatekeeper:

> "Mr. Jones's office."
>
> *"Did you say that this is Mr. Jones's office?"*
>
> "Yes, I did."
>
> *"Who is this?"*
>
> "This is Stacy."
>
> *"Oh, Stacy, this is* (your name) *calling for Mr. Jones. May I speak with him?"*
>
> "And who are you with?"
>
> *"I'm with* (initials of company). *Please tell Mr. Jones I'm holding long distance for him."*
>
> "Is he expecting your call?"
>
> *"I don't believe we have set up a specific time, but please let him know I am on the line."*
>
> "And what is it regarding?"
>
> *"Let Mr. Jones know that I have the answers to the marketing questions. He is supposed to be in, isn't he?"*
>
> "Does he know you?"
>
> *"You know, I don't think we have met face to face, Stacy, but I do have that information for him so please let him know I've called. Okay?"*

Notice several things. In the beginning of the conversation the seller extracted the gatekeeper's name and used it several times in the conversation. People love to hear their own name. Using it helps to soften resistance.

The seller gave the minimum information necessary to answer each question. By allowing the gatekeeper to ask a lot of questions, the seller made the gatekeeper feel that she had done her job properly. At the same time, the seller did not put the conversation in jeopardy by giving the gatekeeper enough information to make a decision not to let the seller through.

The seller tried to respond to each question in the form of a directive, followed by a question designed to get the gatekeeper to acknowledge the directive.

Getting Around the Gate

Sometimes you just cannot get past the gatekeeper. But that does not mean you are helpless—you can still make an end run. Here are some ways to work around the gatekeeper when it is reasonably certain you will not get through.

Direct lines. Try calling the receptionist and asking for the CXO's extension number. If the CXO personally answers that line, you can bypass the gatekeeper altogether. (On the other hand, the gatekeeper may answer that line, too.)

Intentional wrong numbers. Call a different department than the one you want. Let's say you want to call Mr. Jones in Information Systems and cannot seem to get past the gatekeeper. Try calling someone in Accounting and ask for Mr. Jones. That person will tell you that you have reached the wrong department. Then ask that person to transfer you directly to Mr. Jones's office. Often that person can and will.

Call very early in the morning. Often, busy executives arrive at the office by 6 or 7 A.M., long before their gatekeepers show up. So when their telephones ring, there is a very good chance they will answer them directly. This also might get results after work and on weekends.

Call at lunch. At lunchtime, you may be fortunate to have a substitute secretary filling in for the regular one while he or she gets something to eat. The fill-in gatekeeper is usually much easier to get through. Sometimes there is no gatekeeper at all.

Call the office of someone higher up in the organization. If you are trying to reach the executive vice president, for example, then call the chief executive officer's office. The CEO's secretary will inform you that you have reached the wrong office, and may then offer to transfer you to the right party. Of course, a call transferred from the boss's office stands a better chance of getting through.

Cold-Calling via E-Mail

In this day and age you have another communication medium that can be used for cold-calling: cold e-mail. At first glance it looks wonderful: You can send e-mail directly to the desktop computer of the CXO, where only he is likely to read it. No gatekeepers, no phone tag—and no hope for you, either.

There is a word for unsolicited e-mail: spam. The problem with spam (so named because of its lack of good taste, just like a certain spiced meat product) is that it rarely gets to "C" level execs in large accounts. Large corporations filter incoming e-mail, blocking unsolicited mail before you get in. You may sneak through in smaller businesses, but that is not the type of account that will buy your big-ticket items. If you were creative enough to include a file attachment, perhaps a letter in word processing format, the executive's corporate e-mail portal may assume it is a virus and kill your message before the recipient ever sees it. In the end, this might be the best possible outcome, since you don't want the prospect to think of you as "that guy who sends spam."

Instead, you should not use e-mail until you have had some kind of interaction with the recipient. If you have ever traded business cards with him, spoken to him on the phone, been referred by a higher up, or even just attended a presentation that he made, then you have an excuse to move up to e-mail. Refer to that interaction in your first note.

Once you have built rapport with your prospect in the usual way, e-mail can be a valuable communication tool. Not only should you make full use of it to stay in touch, you are expected to do so. But if you open an e-mail channel to someone without justification, you are putting yourself in jeopardy.

What Do Financial Stakeholders Want?

Congratulations, you have finally made direct contact with the Financial Stakeholder. Now you have to say something. As I have emphasized before, you must establish value—but value as it is perceived by the CXO.

A Financial Stakeholder wants you to create "vision value." He is charged with maintaining the vision of the organization and he does not get involved in details. But saying that he does not get involved in details is not the same as saying that he is not involved in the purchasing process; he is involved, both early and late in the process. At an early stage, he determines if your solution supports his vision for the organization's future. He needs to understand current business issues, establish project objectives, and set overall project strategy. Late in the process he approves the spending of the money. In between, he turns over the process to the Technical and Cost Evaluators.

But the CXO is in the market for IT solutions because his vision of the future requires additional technology. He has changes in mind for the organization, so the most important thing you can do when initially communicating with the executive is to present yourself as a "catalyst" for change in the organization. Respect the authority of the CXO but give new perspectives and present valuable information.

Some salespeople get an appointment with the CXO and fail to do their homework. When planning the call, they take no stock in the long-term *value* they can create. Instead they attempt to "sell" this stakeholder on the idea that their solution can reduce costs or increase revenues. But the stakeholder did not grant a meeting to learn how to reduce costs or increase revenue; in his mind, he already knows how to do this. He was hired by the board of directors to do his job based on his proven intelligence and experience, and has devoted hundreds of hours to determining how to reduce costs or increase revenues. That a random seller should walk in off the street and pretend to tell him how to do a basic part of his job is ridiculous and insulting.

During your precious few moments with the Financial Stakeholder, speak with him about the mission, vision, and values of the organization. Show deep respect for the countless hours he and the executive staff have spent planning and executing a well-thought-out business plan. Follow his lead by helping him execute the plan using your solution. Become a conduit for change, an ally of the company, and a supporter of its goals.

If you have a canned sales pitch in mind, keep it in the can. This stakeholder will be listening carefully for indications of whether you are trying to help him or sell him.

First Meeting, First Test

As you might imagine, you will be quickly tested in your first meeting with the Financial Stakeholder. Typically you will have no more than a few minutes to make your case. First impressions are extremely important: Be early, dress like the stakeholder, and speak to him from a business perspective rather than getting caught up in technical details. (I will explain how to do this in coming chapters.) Use the opportunity to raise relevant questions and share business perspectives, hopefully offering new perspectives on the problems facing him.

In your discussions with the stakeholder, demonstrate humility. Humility shows confidence and it is a mark of great leadership. At the appropriate time, pointing out the limitations of your solution will actually increase your credibility with the buyer. The Financial

Stakeholder will also look at your ability to focus, your attention to detail, and your humility. Even executives with the biggest egos respect this trait in other people (when it's combined with confidence). Being a know-it-all and focusing on the solution being sold rather than the buyer's wants is a sure-fire way to be shown to the door.

Meanwhile, keep in mind that some stern questions will be running through the stakeholder's mind:

- Has this seller done his homework (on our industry, our strategies, and our goals)?
- Does he understand our key business drivers?
- Has he been able to convey how his solution or service applies to me?
- Why is his solution or service better than his competitor's?
- Is this individual an empowered decision-maker, or will he have to consult his manager before making decisions?

The Financial Stakeholder will also look for overall peer-to-peer professionalism and for indications of the seller's honesty and integrity. From that he will form a general opinion concerning your "likeness" to him. If he sees a reflection of his own uprightness, he may decide to trust you.

"Like Rank"

Make an effort to match the rank of the buyer whenever possible by bringing along someone of like rank from your company. Meeting executives from a selling company is often important to buying executives. In a joint meeting, the buying executive expects selling executives to know the buyer's business, industry, and current key project, as well as the significance of the relationship in dollars and percentage of business.

This meeting must not be devoted to glad-handing. A "like-rank" sales meeting must reaffirm the seller's commitment and highlight the strategic fit between the two companies. Executives favor suppliers with a similar business philosophy and culture, and prefer to meet with like-rank executives to affirm the value of the relationship.

Guiding Effective Meetings

In a recent online survey from *CIO Magazine,* more than 550 CIOs were asked how they judged the effectiveness of a sales meeting. Their answer was clear: The leading criterion for gauging the effectiveness of a sales meeting was the seller's ability to demonstrate accountability and responsibility. Next came the salesperson's understanding of the customer's business goals, objectives, and challenges. Other behaviors that got high marks in the survey included good listening skills, industry expertise, and knowledge of the executive's business.

A CXO is much more likely to remember a sales meeting that went badly than one that went well. Impressions are formed in the first few minutes of a meeting, and the worst impression a seller can leave is that he wasted the buyer's time. To avoid leaving that impression, take notes during the meeting. Alternately, have one teammate take the notes and another conduct the interview.

Summary

There is no reason to be intimidated by a Financial Stakeholder—if you know how he thinks, are prepared to discuss business issues, and understand what motivates his purchase decisions. Knowing that will allow you to establish trust and credibility, the prerequisites for selling to a CXO. Likewise, gatekeepers can be circumvented, although knowing a few proven tactics helps.

But once you have met with this stakeholder and made a good impression, you still have a long way to go. I will spend the next chapter looking at strategies of how to create SECRET value and examine how to link your solution to the buyer's emotional reasons for buying.

Chapter Five
Deploying the SECRET Weapon: How to Establish Value

LET'S SAY we were to go back in time, before the terms Internet and bricks-and-mortar crept into daily conversation, before acronyms such as AOL, ASP, or CIO existed. A time before sellers and buyers had information pouring over their desktops, and before the alien race of dot.com entrepreneurs and venture capitalists invaded the world. Were salespeople different then?

Not long ago, large, mainstream companies faced with the "technology" age focused on manufacturing and distributing products. They focused on shaping big businesses with large teams of people to do the work. Technology was used to increase productivity or to lower costs.

Today, technology *is* the business. Bricks-and-mortar business today is completely dependent upon technology, from cash registers to forecast reports, from automobiles to airplanes. Companies are changing at every level to meet the efficiencies that technology creates. IT is the business of businesses.

Surprisingly, while the Internet has changed the way that we all do business—more specifically, the way that businesses are organized—it hasn't done much to change salespeople's selling behaviors. Today's technology sellers prospect, ask questions, handle objections, and close much as they did years ago. And today, as they did then, sellers fill out call reports, forecast revenues for management, and are expected at Monday morning meetings. So what is different for salespeople today? It is the way that value is created and the things that buyers value as important that has changed.

Creating value for the buyer has always been a sales enigma. It is a puzzle that is solved by the buyer with the guidance of the seller. Value is critical for establishing long-term sales success, yet it is difficult to clearly understand and define. Establishing (and justifying) value early on in the sales process is a crucial factor in causing a buyer to want to buy. By establishing value sellers define the price they get for the solution, strengthen their ability to compete effectively, and ultimately guarantee their compensation.

Smart sellers spend time preparing ROI (return on investment) calculations for buyers, which shows them how the proposed solution can have economic value. But that's not enough. I've met some salespeople who are amazed at the cost saving their solution will bring, and are even more astounded when the buyer does not move ahead with the purchase. Logically the buyer should commit, but the seller first has to fix the value in the mind of the buyer. And value must come from emotional impact, not just cold logic and cost-based reasoning. In other words, buyers must *feel* they should buy the seller's solution.

Communicating the concept of value can be extremely difficult for the seller. The abstract nature of value makes it difficult to pinpoint. Nonetheless, sellers are compensated for their ability to establish value for the client, and it is essential that they piece together this puzzle.

What Establishes the Value of an IT Solution?

Let's look at some examples of how value is established in the IT market. When the IBM PC came out it was, in terms of technology, a step behind many other desktop systems that were then available. But large corporations rushed to buy it, apparently because of the IBM initials on the box. Without even trying, those three letters established a sense of security. And in a market ruled by the FUD factor (fear, uncertainty, and doubt), the saying in corporate America was that, "No one ever got fired for buying IBM." So the IBM PC sold extremely well to the corporate market.

Within a short time IBM opened retail "centers" to distribute the new personal computer to home users, and launched an aggressive sales effort aimed at small to medium-sized business. It was a huge mistake. The same three initials on the box meant nothing to small business buyers, who were not going to fire themselves for buying something that cost less from another vendor. The only perception that was established was that IBM was the high-price solution. Soon Compaq, Dell, and others overtook the IBM PC, and the rest is history. In hindsight, IBM should have brought the product to the market at a low price-point and captured market share.

If IBM got into trouble for selling reputation instead of price, others have succeeded by appearing to intentionally establish a shortsighted value. For example, when local area network (LAN) technology was first introduced, Novell sold its operating system software, called NetWare, on the basis that it would save the user from having to buy larger hard drives and more printers. Sharing these devices brought about an easy-to-understand economic gain for the buyer. Novell played down the "soft" productivity gains—the savings in time, manpower, and workflow—that might occur, believing that it was too difficult for the client to conceptualize. Today, in a mature LAN market, it is accepted that such "soft" gains are the main reasons for acquiring a network.

There are many examples of the value enigma in the software market. Here is one:

John is pitching software to a mainstream company. The price of software that can support twelve people might be, say, $2,000, but a version of it that can support 100,000 people may cost $2 million. The core technology is the same for both versions, and the user interface is the same, but the value of yet another option—the $250,000 enterprise version—is its ability to *expand* and *scale* to accommodate so many users.

Obviously, John hopes that the buyer will see value in the scalable version of the software. As he walks into the buyer's office, his mind fills with facts and figures about his solution, and he weighs the persuasiveness of each technical feature and its potential benefit. But remember: The buyer is looking for the value in a solution. The sale does not rest on those facts and figures.

The question is: How will John establish the value of his software solution?

How Does the Seller Establish Value?

Unfortunately, each buyer measures value in a different way. In fact, the same types of buyers—ones that share similar responsibilities—will measure the value of a solution differently at any given time. Given two similar buyers in the same marketplace, one may place significant value on one solution while the other may view it as worthless or, worse, a detriment.

Through my research in technology-related sales, I have identified six areas of value that a technology solution can affect (see **Figure 5-1**). The buyer may attach value to one or all of these "buckets" of wants. Some are measurable and some are not, but they can all be significant reasons for the buyer to move ahead in the sale.

The "SECRET" to creating value is:

SECURITY
EXPANDABILITY
COST
RELIABILITY
EASE OF USE
THROUGHPUT

Figure 5-1: The SECRET Values Wheel

We'll spend the rest of this chapter looking at how value is created around these six strategic areas.

SECURITY

EXPANDABILITY

COST

RELIABILITY

EASE OF USE

THROUGHPUT

For many years, security has been very important in IT implementations. With the rise of the Internet, security—physical security, password security, and encryption—is now of paramount importance. According to studies by Forrester Research, the global economic damage caused by security breaches totals more than $15 billion each year. Fearful of such damage, companies are reaching deeper into their pockets to update virus definitions, install firewalls, and set up virtual private networks. International security spending will reach $30.3 billion in the next few years. European firms have been particularly lax in protecting themselves from security threats: More than 50 percent of CIOs and IT directors from across Europe report that security spending accounted for 5 percent or less of their IT budgets.

Security is not a passing fad. Forrester Research released a report asking senior security managers at U.S.–based Global 1000 firms what their security plans would be for the near future. The survey found that firms planned to increase spending on security by 55 percent in the next two years.

Despite all the growth in the security marketplace, many CIOs are still reluctant to acknowledge that they have problems with protecting their networks. They measure security on the basis of whether they have suffered substantial losses, not on whether they are vulnerable to attack by hackers. What is not counted in any of their calculations is the loss from downtime, which can be substantial and should be brought to their attention by the seller.

Another security issue is the concern over privacy, which is reflected in the rise of companies hiring a Chief Privacy Officer (CPO). Until recently, CPOs did not exist, but today they play a key role in the security of the network infrastructure. They address critical questions such as:

- Is our Web site secure?
- Where are the vulnerable points in the network?
- Who has access to which databases?

Most businesses had a person, or people, handling privacy-related issues but the task was typically unpublicized and handled by corporate counsel. Then, a series of security/privacy-related snafus damaged the public perception and stock prices of some companies.

Online retailer Buy.com, for example, accidentally exposed the names and telephone numbers of some customers to other Internet users as a result of a problem with United Parcel Service's Web-based solution return system. Amazon.com was widely panned for its decision to stop guaranteeing that it would no longer share customer information with third parties. And DoubleClick, an Internet advertising company, was forced to cancel plans to merge information about people's surfing practices with personal information on those consumers. In response to such developments, companies such as American Express, AT&T, IBM, and General Motors, among others, are now focused on security as a major IT initiative. Will it create real value to their organizations? Only time will tell.

SECURITY

EXPANDABILITY

COST

RELIABILITY

EASE OF USE

THROUGHPUT

The ability for a technology solution to expand and scale is seen as valuable in any large account. Since many businesses today are growing at breakneck speeds, scalability and expandability are valuable contributions to success.

In the Internet world, Amazon.com began in 1996 as a bookseller, competing with the traditional retail outlets such as Borders and Barnes and Noble.

Now it has grown to sell electronics, garden equipment, music, and kitchen appliances. According to CEO Jeff Bezos, "We (Amazon) will be the largest retailer in the world." The company was able to expand because of the scalability the Internet promises.

Michael Dell of Dell Computer turned the personal computer industry upside down over the past few years by implementing a radical new sales strategy based on direct sales and the Internet. He eliminated traditional channels of distribution, lowered cost of sales by reducing inventories, and then passed the savings to buyers. Dell was thus able to adjust the scale of its business much faster than its competitors. Subsequently, Gateway, Compaq, and others adapted the same sales strategy but with less competitive effectiveness.

Scalability is also important to sellers. In the start-up stage the focus is on a single solution that, when complete, can be delivered through a variety of sales channels, including the Internet. The cost of replicating the solution is insignificant compared to the up-front development costs, thus the solution can be scaled at a continually diminishing cost to a large number of buyers.

I have seen too many implementations that were botched because the solution or technology did not have ample scalability. Given this, what should a seller do if they were an IT buyer and scalability was on the short list? How should their solution respond to this buyer's need?

SECURITY
EXPANDABILITY
COST
RELIABILITY
EASE OF USE
THROUGHPUT

Financial buyers regularly ask, "Is this worth the money?" or "When will we recover the investment?" Financial buyers love to count things: dollars, time, response rates. I have one client who drives her sales team nuts counting things. Her motto is "measuring is managing," and while this does not

motivate the sales force, it is a common behavioral style among financial decision-makers.

The fact is that value can be quantified in dollars, and a good
value appraisal will get you through the most difficult price negotiations. Such an appraisal hinges on several questions such as:

- What will be measured, and when?
- How will this solution save money?
- How will the implementation save manpower costs?
- Will this implementation increase revenue?

When someone faces a decision to buy, he weighs the value of the
solution against the cost to implement it. When the value outweighs the
cost, he is likely to buy. In a small sale, reasons other than the ones
described in the SECRET can cause buyers to move ahead. Superficial or
frivolous wants can replace cost justification. In a large sale, your ability to
link the SECRET values to the cost justification is critical, in turn developing a real and compelling Return on Investment (ROI) (see **Figure 5-2**).

Figure 5-2: Value Versus Cost Justification

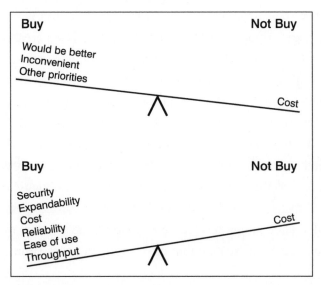

An industry publication, *CIO Magazine,* surveyed 256 CIOs on how they measured IT value. In terms of company size, 39 percent were employed in organizations with 5,000 employees or more; 22 percent in organizations with 1,000 to 4,999 employees; and 38 percent in organizations with fewer than 1,000 employees. One fifth worked in computer-related industries. Here's what they said was most important:

- Return on Investment (41 percent)
- Total cost of ownership (29 percent)
- Internal rate of return (14 percent)

When asked which factors were included in value equations, the CIOs listed:

- Costs and expenses (85 percent)
- Productivity (67 percent)
- External customer satisfaction (55 percent)
- Soft benefits (48 percent)
- Uptime (46 percent)
- Revenue (45 percent)

There being no one way to establish value, it is very important for the seller to find out what will be measured when creating value. I always ask, "How will you know when my company has done a good job?" Sometimes buyers will have no answer, either because they have never thought of it or because they don't want to say. Ask questions that are planned and rehearsed, to the point that they sound natural and conversational to the buyer. Establishing business rapport and aligning with the buyer will help any seller get an answer faster.

Acknowledge, too, that there may not be a ready answer because many IT managers avoid the excruciating exercise of assessing value. The value of acquiring new equipment cannot be assessed by one neat calculation, but may require four or five different ones. Many managers also do not have a clear picture of how assessing value affects day-to-day operations.

There are many times in the life of an IT-selling organization when knowing the value of assets is crucial, including:

- When submitting funding proposals to senior management
- When it is time to replace or upgrade equipment
- When a major change in business operations calls for a change in information technology
- When assessing ways to increase competitive advantage
- When considering an outsourcing proposal
- When helping a buyer to assess the value of equipment in order to justify the purchase

IT Financial Stakeholders usually calculate value beginning with the acquisition cost: What was paid for the equipment two or three years ago? They also have to figure out what they could get for the equipment if they tried to sell it now, and what it would cost them to buy new equipment to replace its function. There are also less quantifiable considerations: Does this item enhance work efficiency, customer satisfaction, or business practices?

To find an item's current worth, you must determine the market (or fair market) value; in other words, what one could sell it for in its current condition. There is a range of prices, depending on the condition of the equipment. Find out market value by checking with manufacturers that have price lists for buybacks. Also, refer to trade publications and the Internet for classified ads offering similar equipment.

There is also book value to consider, which assumes a predetermined amount of wear and tear over the years and is similar to market value. If you've taken especially good care of your equipment, however, its market value might exceed book value. On the other hand, if you've abused and battered your equipment, you might not get the full book value on the open market.

It is important for the seller to keep in mind that, when asked about the reliability of IT value measurements, only 10 percent of the CIOs surveyed believed value measures were "very" or "completely" reliable. Fifty-five percent indicated that measurements were "somewhat" reliable, while 30 percent indicated that they were "not reliable." Apparently, while using cost calculations is a logical exercise, the buying community in most sales situations does not trust them. Therefore, sellers should not spend a lot of time on them unless they are sure that the buyer is among the 10 percent who do trust them.

The term "time is money" is surely felt among IT staff. When a large system fails, heads roll.

E-Trade, the leader in online trading services, experienced massive traffic causing its global network to falter. As a result, its stock fell more than 30 percent in one year after complaints surfaced about its inability to execute trades during heavy market activity. The straw that broke the camel's back came when the market experienced a 550-point drop in the Dow Jones Industrial Average one day and a 330-point rally the next day. The resulting heavy usage eroded reliability, and E-Trade's customers took the firm to court. A class action lawsuit was filed to prevent E-Trade from taking on additional accounts until it was able to guarantee timely access to accounts and transactions for its existing customer base. The customers were also looking for monetary compensation for gains lost when the company was unable to execute specified trades.

Recently an article in the *Washington Post* told of 50,000 Verizon customers who were left without e-mail; the e-mail simply "disappeared." In this and similar incidents, blame was placed on "difficulties installing and maintaining its high-speed Internet services." That's a euphemism for poor reliability.

All of us have experienced network breakdowns and been frustrated with the time wasted and money spent fixing an IT implementation that was poorly executed. A seller should be ready to answer these important questions: How will your solution improve reliability? And do you have existing buyers who can attest to this reliability?

Growing dependence on fast-paced electronic interaction—among customers, employees, and trading partners—has left little tolerance for downtime in computing environments, scheduled or otherwise. High-availability computing, which allows for reliable system uptime and was once considered a strategic advantage, has now become a tactical necessity. In response, CIOs are scrambling for technological solutions that can deliver a high level of service at a reasonable cost.

When considering service levels, how does a CIO decide how high is high enough? For some companies, the goal is less than an hour of unplanned downtime a year. For retailer Sears, Roebuck and Co., computer operations are expected to work 24/7. For United Airlines, the uptime target for "mission-critical systems" is 99.97 percent, which is almost as high as "business-critical systems" that use 99.7 percent as the target. The first is less than three hours a year; the second is about two hours a month.

For retail drugstore chain Rite Aid, the normal maintenance window for its centralized drug interaction system narrowed with the company's nationwide expansion. With other enterprises that fill mail orders around the clock, such as L.L. Bean, any time the computer is down, orders do not get shipped and that hurts profits. The real question would appear to be: What is the threshold of pain that an account can tolerate?

I predict that within five years high reliability will become more of a commodity, as vendors will be forced to embed reliability features in the personal computers and servers they offer. In the meantime, reliability is a major technology concern for IT buyers, and sellers (and their solutions and services) will be tested.

Imagine a computer with no Windows operating system; the very first computers had no operating systems at all. The scientists and mathematicians who were experimenting with computers did all of their own programming, data entry, computer operation, and repairs. Because these early computers cost millions of dollars, it was considered of utmost importance to keep the computers in continual use. So the select few who were permitted access to these early computers would sign up for time at all hours of the day or night. During their allotted period, they would have complete control of the computer, loading up their software and data and doing their work. The users *were* the operating system. But in the 1960s, as use of computers shifted to mainstream businesses, they were operated by laypeople and operating systems became more and more important.

This is what lies at the heart of the ease-of-use issue: computers adapting to the operator rather than forcing the operator to adapt to the computer. A perfect example is Apple Computer. Its Macintosh operating system revolutionized computing. For the first time, an operating system and its application software recognized that the human operator was more expensive than the computer system. Since then, just about every operating system has adopted the graphic user interface. Companies such as Microsoft have become powerful as a result of successfully marketing Windows, a poor answer to the Macintosh operating system.

Here is another example from the world of telecommunications. In 1988, two Stanford University graduates, Leonard Bosack and

Sandy Lerner, invented the super-fast network technology called the "router." This innovative hardware device allowed previously incompatible software protocols to communicate at blazing speeds with relatively simple plug-and-play installation. It changed the way the world viewed computer networks and led to the birth of Cisco Systems.

But it was how it addressed ease-of-use issues that gave Cisco a buying value advantage. Cisco found that IT buyers were confused by the solution strategies of the competition and the standards being developed. Facing large procurements and long-term life cycles, buyers were unhappy with vendors' aloofness, and looked to Cisco for leadership. John Chambers, Cisco's CEO, set up focus groups with large customers and asked what he could do to help. He discovered that buyers were concerned about finger-pointing among uncooperative vendors. "For each dollar they spend on networking hardware and software," Chambers later said, "they would spend three or four dollars integrating and administering them." The solution was to change Cisco into a one-stop shop for wide-area network (WAN) services. Chambers went to the market to acquire companies to provide the complete solution. Cisco is now grossing billions yearly and dominating the WAN connectivity switch/router market.

Ease of use will always be a primary concern for IT buyers and must be addressed up front by the seller.

SECURITY
EXPANDABILITY
COST
RELIABILITY
EASE OF USE
THROUGHPUT

In the IT world, "throughput" is known as the amount of work that can be performed by a computer system or component in a given period of time. If you believe throughput is a technical rather than a business issue, consider the experience of America Online (AOL). AOL experienced a throughput-related business crisis that could have been catastrophic. When the company went from a pay-per-use to a monthly flat rate pricing plan—turning what AOL considered a cash cow into a loss leader—membership sky-rocketed from six million to well more than eight million in less than twelve months. But the result was cyber gridlock: busy signals and upset subscribers. At one point, eight million AOL members were trying to dial in to only 200,000 modems, a ratio of forty subscribers per modem, when a twelve-to-one ratio is considered optimal. Customers trying to log on to AOL during peak evening hours were unsuccessful about one third of the time, a long way from the industry average of a 9.5 percent failure rate. In response, AOL invested $350 million in system upgrades. It also invested millions more in reimbursing customers who sued for loss of service.

Dell Computer represents another throughput story. Not long ago Dell was in third place among personal computer vendors. That wasn't good enough. After poring over projected forecasts and holding many high-level strategic planning discussions, Dell executives decided to change the way the company manufactured computers.

Dell had in place a fairly standard assembly-line process for making servers and personal computers. Computer cases would roll along with one person after another adding a component; it took up to twenty-five people to build one machine. Dell's teams designed a method for quickly moving components into the plant. Then they had to move the completed machines out as fast as they were finished. They also had to find a way for the manufacturing teams to know when they had all the components needed to finish an order—whether it was for a single computer or for two hundred of them—before they started building the first one. That simple throughput change led Dell to crank out one million machines every three months as it climbed past Compaq to become the top seller of personal computers in the United States.

Summary

While the SECRET will let you probe the areas most likely to be of interest to an IT buyer, what matters most is the intensity of how the buyer feels about the SECRET topics. He may be indifferent to some or most, but in any successful sale there will be at least one or more SECRET about which the buyer has experienced frustration and can be led to a decision to buy your solution. In the next chapter, we will examine how to specifically link the SECRET to your solution. This is a step-by-step approach leading the buyer to the seller without making them feel pressured. This approach takes theory and puts it into practice—and doing that is a bigger secret than the SECRET itself!

Resource File: The SECRET Worksheet

Using the worksheet on page 88 (see **Figure 5-3**), fill in the areas that you feel the buyer would most value. Use the questions in each area to guide your team discussions.

Security

- Which of these areas are of concern?
  - Physical
  - Password
  - Encryption
- What problems have been encountered?
- What losses have occurred?
- What anticipated changes will require security enhancements?
- What are the areas of vulnerability?
- Who has access?

Expandability

- Can the present system meet future growth?
- How is the present system limiting operation?
- Is it interfering with productivity so revenues are impacted?
- Is the present system scalable to accommodate future growth?
- What anticipated expansion in the future will render the present system obsolete?

Cost

- Is the initial cost of the system justified as an investment?
- Has the initial cost of the present system been an investment for the company?
- Has ROI actually been calculated?
- Has the cost of operation been what was anticipated?
- Are there areas of concern with the present system?
- Have service and support costs exceeded expectations?
- Is the present system giving the anticipated returns in terms of productivity?
- Has downtime affected productivity?
- How often has the system been down?
- Has the cost of downtime been calculated?
- In what areas could revenue be increased? Give reasons.

Reliability

- What integrity does the present system have?
- How important is reliability to you?
- Is your data time-sensitive?
- Has the system supported time-sensitive data?
- Have there been breakdowns in the system?
- How often have you had to call for support?
- Has the system performed as you anticipated?

- If not, what have been the problem areas?
- Has the present vendor been timely in resolving these problems?
- Has your team had to service the system themselves?
- Has productivity been affected?
- In what ways?
- What are overtime costs?
- Has lost productivity interfered with your buyer relationships?

Ease of Use

- How easy is the system to use daily, and for new employees?
- How long does training take until a user is up to full productivity?
- Have the users had complaints?
- In what areas?
- Are they justified?
- Has this interfered with productivity?
- What would the users like to see changed?

Throughput

- Has throughput of information affected productivity?
- Have there been any problems with throughput?
- Have they caused any buyer complaints?
- Have they interfered with productivity?
- How has that affected cost and time?

Figure 5-3: The SECRET Values Worksheet

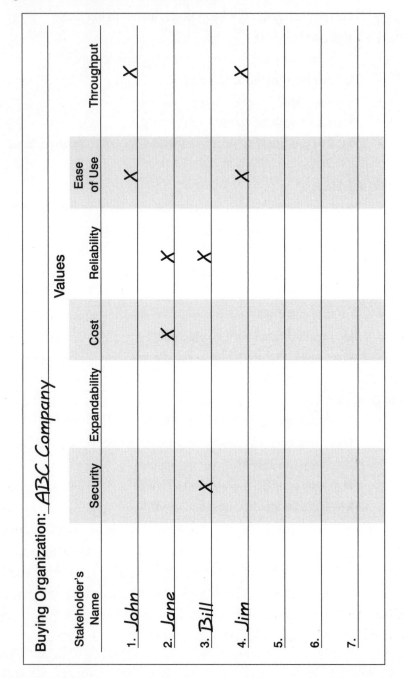

Buying Organization: _ABC Company_

Values

Stakeholder's Name	Security	Expandability	Cost	Reliability	Ease of Use	Throughput
1. John					X	X
2. Jane			X	X		
3. Bill	X			X		
4. Jim					X	X
5.						
6.						
7.						

Chapter Six
Closing the GAP: Getting Commitment in the Major Sale

IN CHAPTER 5, I suggested that success in creating value depends upon understanding the behaviors of the buyer and relating those behaviors to the SECRET. Those sellers who are the most effective at identifying and creating value are most likely to be the top performers. Not surprisingly, the top performers are also the most skilled in asking effective questions that flesh out the buyer's SECRET values. So let's devote this chapter to exploring the most underrated skill of all—asking questions.

Let's say you have banged your elbow and have gone to see a doctor. As you enter the waiting room, you approach the receptionist to fill out forms in order to give the doctor the necessary information on the medical problem and for billing purposes. When you get in to see the doctor, he examines the gathered information and begins feeling around the arm, gauging your reaction to his touch. There's not much reaction along the wrist and forearm, but at the elbow you flinch. He nods and orders x-rays to be taken. Satisfied with what he sees, the doctor puts a splint on your arm, prescribes a painkiller, and schedules a follow-up visit. Satisfied with the treatment, you leave the office for home.

This is a plausible scenario. But say it happened a completely different way. What if, before you said anything about the elbow, the doctor had said, "It seems to me that you need a hair transplant." Or he had said that your real problem is that you fell because you were overweight, after which he tried to enroll you in a weight loss clinic. What if he had fit the splint but then tried to sell you a nose job?

Actually, you might be overweight, balding, and have a crooked nose, but that's not why you visited the doctor. What motivated you to visit the doctor was the pain from the banged elbow. You had no emotional involvement in those other issues. The doctor had a low likelihood of "selling" you those other services because they were not of high enough interest to you at that moment. If he *had* tried to recommend those additional services, there was a high likelihood that you would have selected a different doctor the next time. Or worse, if you had actually purchased these services, you could have sued him. As the saying goes: "Prescription without diagnosis is malpractice."

The parallel with the IT sale is obvious, the seller being the doctor and the buyer, the patient. Fortunately, the prospects typically are not bleeding when you see them, but they are feeling some sort of pain. Your challenge is to have your buyer accurately describe that pain, to diagnose the problem, and prescribe a solution.

Implied Versus Stated Wants

Traditionally sellers are taught to sell to buyers' needs. In recent years, selling to needs has become less attractive to buyers, as they have become more educated and sophisticated in their buying prowess. In truth, IT buyers no longer live in a "needs-based" society. And because of the affluence of buyers in corporate accounts, they have very little need for anything, especially not for the solution you are providing. Getting buyers to *want* your solution is by far a more powerful sales approach. It creates an internal pressure to buy, deliberately getting buyers emotionally involved as you create a bond of trust. But before the buyer acts on those wants, the seller must learn to distinguish between implied and stated wants.

Implied wants are those that are not owned by the buyer. Many times, when a buyer explains to the seller what he wants, he frequently ties the word "should" to it. For instance, "My team wants to implement the project, and I think they should," or "I should move ahead on this proposal," or "We should invest in a faster network," are examples of implied wants.

A key to guiding buyers toward a solution is to get them to state the want in no uncertain terms. Stated wants bring ownership to the direct SECRET value of your solution. Statements such as, "I want this solution to provide 24/7 uptime," or "I want absolutely no downtime," signal strong feeling, allowing the seller to link value to the buyer's desires.

Figure 6-1: Implied Versus Stated Wants

<u>**Implied Wants**</u>

My team has indicated . . .

I think the problem is . . .

I've found that . . .

<u>**Stated Wants**</u>

I want my team to . . .

I want to fix . . .

I want the problem to be . . .

In truth, implied wants do not have the emotional pull of stated wants (see **Figure 6-1**). Stated wants evoke intangible, emotionally charged feelings. When wants are clearly stated by the buyer, he is stating a willingness to take action for his own reasons. Stated wants have the power to illuminate emotional motives as much as car headlights illuminate the dark road ahead. Implied wants, by contrast, are more like flashlights. Stakeholders are more likely to take action on your ideas when they feel a strong emotional connection to them. The seller's job is to illuminate their emotions so they fully understand the relationship between smart decisions and fulfilling stated wants.

When you engage in conversation with stakeholders about their stated wants and they become emotionally involved, you can probe the way they rank the wants, and then look at each in turn. The next step is to apply a want to a SECRET, which makes a value connection for the buyer. This attaches the buyer to you, helping you understand them at a deeper level and dramatically shortening the time it takes to gain their trust. Emotional involvement and trust are two critical elements in creating peer level relationships with buyers. Because a stated want involves emotional commitment, the buyer has made a decision that turns the desire into a goal. The moment he changes gears from implied wants to stated wants, you can see dramatic changes in his face, posture, and tone. The subject is alive to him and he is ready to hear how the want can be related to what he sees as valuable—the SECRET.

The GAP Model

The GAP questioning model is designed to help you identify implied wants and turn them into stated wants. It was developed by asking sellers the question, "What do highly successful sellers do that unsuccessful ones don't?" The answer: They ask lots of questions. By asking questions, they engage buyers to talk about their problems and guide

buyers toward creating their own reasons to buy and develop ownership of a solution.

Effective sellers have an almost innate ability to gain control of
the selling situation, creating instant rapport with the buyer. Their
conversational style is relaxed and casual with even the most difficult
buyers. With this "conversational" style of asking questions, the prospect
becomes an active participant in the solution, or the "prescription."

Further, by asking questions and linking the answers to the value
SECRET, objections are handled by prevention, and closing is an easier
task. Instead of pressuring the buyer to make a decision, the buyer takes
ownership for the sale and creates his own internal reasons to buy.

It is clear that successful sellers ask effective questions in a particular order.

Figure 6-2: The GAP Questioning Model

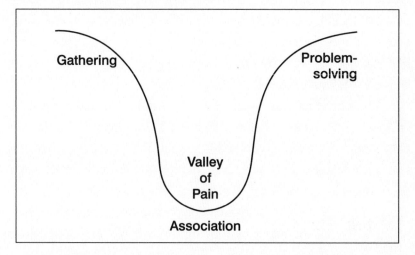

- **G for Gathering.** Questions that invite factual responses from
  the buyer while gathering information about the buyer's current problems.

- **A for Association.** Questions uncovering emotional reasons to buy, effectively linking the SECRET to your solution.
- **P for Problem-solving.** Questions that help the buyer solve the problem by creating "ownership" for a solution.

Posing questions helps develop the buyer's involvement. The less you say, the better. When the prospect talks, his entire focus is on what he is saying. When you talk, the process is less engaging for the stakeholder and you could lose his focus. Stakeholders tend to be able to clarify their thoughts by talking rather than listening. On the other hand, when you are the one doing the talking, you will lose up to two thirds of their focus and attention.

Also, if you do too much talking, you can get caught in the trap of solving the problem before the buyer is emotionally involved in the solution. Engaging the emotions of the stakeholder is critical to moving him forward. Many inexperienced sellers jump from G to P in the GAP model, completely ignoring the emotional aspect, thus failing to engage the stakeholder emotionally. If you do, you'll come across like that doctor, who tried to diagnose the problem without probing, and as a result prescribed the wrong treatment.

You must first draw emotional "associations" to find out what the prospect really wants. After all, the buyer knows more about the company's business than you do and is caught up in the day-to-day emotions of implementation. Whatever solution you prescribe must engage the prospect at an emotional level before he will associate any SECRET value with it. Unlike our blinkered doctor, you must determine what is really bothering the buyer by moving through the Valley of Pain as shown in **Figure 6-2**. The valley is filled with SECRET associations to wants that trigger reasons to buy.

When you walked into the doctor's office, you may have mentioned a desire to weigh less, but that did not signal a desire to be sold weight-loss products. In the final analysis, you walked in desiring many things, as we all do all the time. Losing weight was one desire that you may have implied. But your *stated* desire was for help with your injured elbow. The elbow pain was your emotional focus, even if you had other pains as well. The doctor will not know this without consulting the patient. It is the patient's perception of the "pain" that the doctor must understand before prescribing.

What an experienced seller does, therefore, is to ask gathering questions to elicit information that the seller could not find in the earlier research phase before the meeting. Then, the seller asks emotionally charged questions to reveal the buyer's wants. As he does that he listens for the implied wants versus the stated wants. He uses these associative questions to attempt to turn implied into stated wants and then to link those wants to the SECRET, one by one. The seller can then ask questions to help the stakeholder own the solution and thereby solve the problem himself.

Let's go through each category of questions in turn.

"Gathering" Questions

Gathering questions are those that gather the necessary information to help determine the buyer's current situation. These are in addition to the usual who, what, where, when, and why: How many do you have? How many do you need? How much RAM does it have? What's the throughput?

There are five questions that you should ask as early in the process as possible. Answering these will move the sales process forward quickly, and without them the sale will slow down or even derail.

1. What is your main objective? (And, how will you know when you've achieved it? If you had three outcomes for this project that would make it a success, what would they be? What would you need to make everything perfect?)
2. Have you allocated funds for this project? If so, how much?

3. Who, beside yourself, will make the decision to move ahead?

4. Are you working against a particular deadline?

5. What are your decision criteria? (Is a proposal necessary? What is the buying process? Who is involved?)

Other questions you may want to ask:

- Who else should I talk to? (And why are they important?)
- What is the nature of the project/request? (Who cares about it? Who is watching it? Are there any company-wide implications?)
- What is the project's name?
- Who are its influencers?
- What major considerations stand in the way of what you want?
- How are you currently dealing with the problem that the project addresses?
- What other problems are you experiencing?
- What are you using now? (What do you like most about it? What do you like least about it? What alternatives have you considered? When did you consider them?)
- Do you spend a lot of time maintaining systems?
- Do you plan to do a business process re-engineering?
- How do you ensure consistency in your applications?
- How may we help?

These are not scripts: The precise wording is not important. Instead, these are topics that you should explore in a natural conversation with the buyer. If it seems like a long list of topics, just keep these four themes in mind:

1. What do they need?
2. What is the budget?
3. Who has the decision-making power?
4. What is the deadline?

Get these questions answered when and where you can, and do it as soon as possible in the process.

Gathering from Financial Stakeholders

The Financial Stakeholder is by far the most important stakeholder. As discussed in Chapter 3, preparation is critical when approaching this person. Your questions should be rehearsed and show a deep understanding of the organization. To avoid annoying him, ask the minimum number of "gathering" questions. Keep questions focused on current events and ask questions that only he can answer. Remember: Vision, mission, and values are the focus of these stakeholders.

With the Financial Stakeholder, the key is to focus on benefits (how the features apply to the long-term wants) as they relate to the company's big-picture situation, while remaining sensitive about relationships.

Questions might include:

- What are the specific goals for the company over the next month, quarter, year?
- What are the criteria for establishing a relationship with the company?
- What are your key objectives?
- What is your key competitive advantage today?
- What is your vision of the implementation?
- What is most important to you?

Gathering from Champion Stakeholders

Champion Stakeholders can be the source of many answers that you were not able to obtain elsewhere. When talking to the Champion Stakeholder, focus on the advantages of your product, and especially describe how its features will improve the life of the stakeholder.

Questions that you may want to specifically address to the Champion Stakeholder include:

- What are the departmental needs?
- What are the current and future projects and priorities?
- How long has the organization used the current supplier?
- What is the stakeholder's greatest concern?
- What is the stakeholder's perception of our organization?
- What is your knowledge or experience with stakeholder X?

Gathering from Technical Evaluator Stakeholders

The Technical Evaluator Stakeholder is easy to talk to, but you must focus on "gathering" relevant information rather than gossiping about industry trends. Technical Evaluators are interested in features (the factual characteristics of the solutions offered) as well as functionality and detail. They can be very specific about how your solution should interface with their problem.

When working with a Technical Evaluator, get as many questions answered as soon as possible. While they are easy to talk to, they do not want you talking to anyone else. Once they come to know you, they see you as a source of information, and information is the lifeblood of their job. They will see no reason to share you with anyone else, and that means the answers you get from them will be increasingly misleading as they string you along in hopes that you will keep coming back, effectively blocking your path to other buyers in the organization.

While they typically cannot say "yes," they can veto your solution based on their technical perception and bias. Because other stakeholders rely on their technical evaluation as part of their own decision process, it is essential to have them "buy in."

Questions to ask include:

- What are the technical criteria?
- How much?
- How big?
- How many?
- How often?

Gathering from Cost Evaluator Stakeholders

Sometimes called the purchasing agent in a major account or the contract officer in a government account, this stakeholder is a key player. Your experience with this person is likely to be similar to the Technical Evaluator, except you are likely to have less contact with the Cost Evaluator.

A tendency for inexperienced sellers is to penetrate an account at this level. For experienced sellers it is just the opposite: They tend to ignore the Cost Stakeholder. Neither approach works, as this person is key to the process and should not be forgotten. Get Cost Evaluators in the loop early and keep them informed as you move from stakeholder to stakeholder. As you speak with them, keep the questions focused on cost issues and vendor paperwork requirements. Again, get critical questions answered as early as possible.

Questions for the Cost Evaluator include:

- Who controls the budget for this project?
- How can we expedite the buying process?
- Is all the paperwork in?
- Who needs to sign off on this?

Gathering from End User Stakeholders

End users can be a source of background information. When talking to users, focus on how the solution will improve the work environment.

Questions that you might want to specifically ask users include:

- What are you using now?
- How are you using it?
- Who is who in the department/company?
- What does this company make/do/deliver?
- What part of the process are you involved in?
- How could that be more efficient, effective, secure, reliable, or cost effective?

"Association" Questions

Association questions are those that link the buyer to the value. Most of the time, in the gathering phase, the buyer implies wants. The purpose of association questions is to allow the buyer to state their wants. It is up to you to get the buyer to restate implied wants as stated wants. Association questions, therefore, are emotionally charged questions that will cause him to talk about the effects, implications, consequences, or problems that will reveal what he wants.

As the behavioral scientist Abraham Maslow described in his early research, people will move faster away from "pain" or "trouble" than they will toward "pleasure" or "growth." So phrasing your question to illuminate pain or trouble is usually more effective. That is why I call these "association" questions—associating the SECRET value to the pain or pleasure they will receive in buying your solution (see **Figure 6-3**).

Keep in mind, however, the difference between risk and pain. Risk is something the prospect likely associates with pain, since he already has enough uncertainties. For sellers, this creates a challenge, since sellers are often types who embrace risk, seeing it as a source of excitement or possibilities. So they will ask risk-related questions, as if the project was a mutual adventure. They may attempt to motivate the buyer with pleasure-inducing questions such as, "Wouldn't it be great to be able to implement that solution?" or, "How will the team feel about this?" These may work for some buyers, but for most they bring fear into the hearts of the listeners.

Figure 6-3: Pain/Pleasure Principle

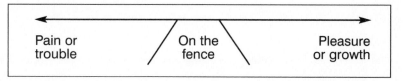

Instead, the seller needs to ask pain-related questions such as:

- What would it cost, ultimately, if things stayed the same?
- How many end users/resources has this problem affected?
- What do end users say to you when they have problems?
- What is the impact of . . . ?
- What are the consequences of . . . ?
- What would happen if . . . ?

When exploring SECRET topics with association questions, always fully probe into one topic at a time in an attempt to have the buyer state a want. Remember: Sellers create value by linking the SECRET to stated wants.

Stakeholders may try to pressure you into giving a solution before you have the chance to take them through the association phase. "I've told you what I know. What do you have to offer?" they will say. Be careful not to present your solution before the link is made to the SECRET. Otherwise, you are doing a disservice to both the buyer and yourself, since not asking association questions lowers the likelihood of having them buy. It's better to tell the buyer, in a straightforward way, that you don't have the information you need to recommend a solution, and ask permission to pose more questions.

When you try to make a sale on an implied want, you put yourself in a precarious situation, since the stakeholder has no ownership of the solution and no motivation to take action. You also risk alienating the stakeholder, since he will feel that you are pushing him into doing something that he is at best lukewarm about doing. The prospect's perception is that you are only interested in "selling" your product or service, not "partnering" to prescribe what is best.

Once you arrive at the association phase, you may feel uncomfortable asking emotionally probing questions. But remember: The more immediate and raw the wound, the more likely it is that the prospect will act to bring about a solution. But even then, *if you try to jump from G to P, your prospect will have little motivation to use your solution,* and you may find yourself trying to supply that motivation with a flood of features and other meaningless technical descriptions.

Figure 6-4: Jumping the GAP

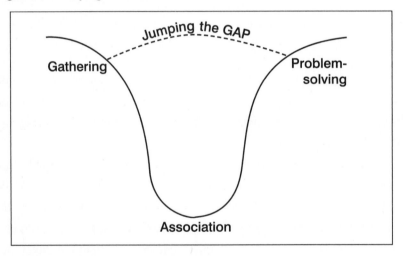

"Problem-solving" Questions

Problem-solving questions are those that help buyers discover how they will benefit from the solution and encourage them to own the idea. These questions must be asked at the appropriate time in the selling process. When asked at the right time, they can powerfully illuminate the situation so as to reveal your solution.

At this point you and the stakeholder have defined the problem with association questions, and through these questions the buyer has stated an emotional attachment to it. He is now inferring that he wants a solution to this problem.

Since you know the solution better than the buyer, it's easier for you to see the solution. Be careful, at this point in the conversation, not to solve the problem for the buyer with a presentation. The purpose of these questions is to have the buyer put the spotlight on the solution, enabling the buyer to discover the solution and take ownership of it.

Solution sales are complex and, as complexity grows, finding the "right" solutions is not always easy. Your solution may be part of a larger picture and affect many areas of operation. So there may not be a straightforward relationship between your solution and the buyer's stated wants. If this happens, the buyer may see your solution as risky, so tread carefully.

Here are some samples of problem-solving questions:

- In the best-case scenario, how would you solve this problem?
- How would you see this problem being improved?
- What if I could show you a way to . . . ?
- How do you plan to achieve that goal?
- Why is it important to solve this problem?
- Is there anything I've overlooked?
- What do you see as the next step?
- How soon would you like to start?
- In a perfect world, if you could have things any way you wanted, what would you change?

Sellers may ask only one or two problem-solving questions in a conversation with a buyer. When asked at the right time, in the right way, they are extremely effective.

GAP in Action

What would the GAP model look like on a sales call? First, keep in mind that GAP is a "model" not a "process," so asking questions should have a conversational, unrehearsed flow. You may ask gathering, association, and problem-solving questions in any order that feels right. It is dynamic in nature and should always be matched to the behavioral style of the buyer as discussed in Chapter 3.

The way the call begins will depend on who asked for the meeting. If the stakeholder called you, you should have a good idea from that request where his interest lies, since you probably asked some gathering questions at that time. If you called the meeting, you may know less—unless the Champion Stakeholder has been very forthcoming with specific information—and will want to ask more of these questions at the meeting.

Let's say that you called the meeting. You have begun the conversation with some rapport building and are ready to get down to business. Here are some questions you might ask the buyer:

- How many users do you have in the division?
- How long has that system been in place?
- Is the system supported "in-house" or by a provider?
- Is the same system used in other divisions?
- Are you looking to replace the system or upgrade it?
- Is there a budget to support that plan?
- Who beside yourself would be involved in the decisions to do so?

At this point you are ready to wrap up the gathering phase by asking questions that will lead you to the association phase:

- What do you think are the advantages of the present system?
- What would you improve about it?
- What do you think are its weaknesses?
- What kinds of problems have you had with it?

These last questions are transitions that lead to the emotionally charged association questions that motivate the stakeholder to want to buy your solution. Each question should logically be based on the stakeholder's reply to the previous question. For example, in reply to that last question, the stakeholder may say:

Buyer: "We have had some downtime problems. The staff produces time-sensitive data, so this can be something of a problem."

Seller: "How often has that happened?" (Gathering)

Buyer: "Well, maybe twice a month."

Seller: "For long periods?" (Gathering)

Buyer: "Actually, several times, for as much as six hours."

Seller: "How did that impact your budget?" (Associative)

Buyer: "It is always costly to have staff being unproductive. The remote offices are dependent on the data so it affects them as well." (Cost is a SECRET)

Seller: "That sounds like it's affecting profits." (Associative)

Buyer: "It is, I'm sure. With data needs like ours, we need reliability." (Stated want)

Seller: "Don't you have a service contract with the present provider?" (Gathering)

Buyer: "We do, but they are not very responsive. Our staff is busy 24/7. When they can't get customer support to respond in the middle of the night or on weekends, the system gets bogged down. Then I get calls, regardless of the time, day or night. We could use a system that is more reliable and technical support that we can count on!" (Implied want)

The stakeholder has attached emotion to the issues of cost and reliability. With reliability he articulated a stated want (reliability is the R in the SECRET topics), but what does reliability actually mean to the customer? To find out, you ask another associative question linking your solution to the stated want.

Seller: *"Mr. Stakeholder, what does reliability actually mean to you?"* (Associative)

Buyer: "Actually, for our company that would be customer support 24/7 just as our production needs dictate." (Stated want)

Now that you have developed a stated want, and a very specific one at that, the buyer has clarified what he means by reliability. It's time to link your solution with a problem-solving question.

Presenting the Solution

Okay, you've gathered your facts and made your associations. The buyer has stated his wants and you've linked them to the SECRET. It's time to present your solution—to present *the* solution.

What I mean is that you have to bind the presentation in a way that shows that the buyer's stated wants, and your solution, are interdependent, and that together they create value for the buyer. Conclude with a statement that shows the value of your solution by tying its benefits to the stakeholder's emotional reasons to buy. It should include:

- A specific feature or characteristic of your solution that relates to the stakeholder's stated needs
- An appeal to the stakeholder's emotional investment

- Examples of successful implementation of the solution
- A challenge to act, for the customer

When he does act on it, it should bring about the next step of the sales process (see **Figure 6-5**).

Figure 6-5: Presenting the Solution

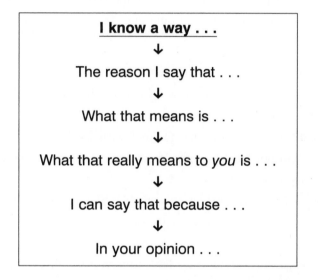

I know a way . . .
↓
The reason I say that . . .
↓
What that means is . . .
↓
What that really means to *you* is . . .
↓
I can say that because . . .
↓
In your opinion . . .

A conversation with the buyer should go something like this six-step process:

1. Present your claim that you have a solution. "I know a way you can . . ."
2. Refer to one operative feature of your proposed solution. "The reason I say that is . . ."
3. State the advantage offered by your solution. "What that means is . . ."
4. Point out the personal benefit offered to the stakeholder by your solution. "What that really means to you (use stakeholder's name) is . . ."

5. Point to past examples, or other evidence. "I can say that because . . ."
6. Test commitment. "In your opinion . . ."

Let's go back to the customer who has been plagued with complaints from the staff at all hours concerning technical issues, since the organization began operating 24/7. To meet the buyer's wants, you are proposing a 24/7 maintenance contract. Your closing may go something like this:

"I know a way to solve your uptime issues."

"The reason I say that is we have implemented 24/7 customer support programs for our system users."

"What that means is that anyone with 24/7 production in your company will not have to suffer extended downtime, resulting in lost productivity and revenue."

"What that really means to you, Mr. Stakeholder, is that no one will be calling you at 2 A.M. with technical problems. They will be calling us."

"I can say that because we have implemented this service support at Marriott Corporation (give real buyers with recognizable names that are in the same industry or production mode) for six months with absolutely no complaints."

"In your opinion, Mr. Stakeholder, would this solve your customer support problem?" (If the customer gives a conditional "yes" or even a "no, that's not quite what I meant," then you ask the customer to clarify his stated want and repeat the presentation process. With a yes, you ask to move to the next step of the process. That may be going to another stakeholder or a committee.)

You have now closed the GAP between the buyer's wants and your solution.

Summary

Successful sellers know that mastering the art of asking emotionally charged questions in a conversational way is an important ingredient to success. Uncovering stated wants and then linking them to the SECRET values through presentation statements shortens the buying process. While these personal skills are important, they are not enough. Having arrived at this point, you have interacted with key stakeholders. We now need to look at managing the proposal process and then managing the buyer after he is engaged in buying. We will do that in the next chapters.

Resource File: GAP Questioning Worksheet

This worksheet has two phases. The first is the development of questions designed around the value that your solution has to offer. This requires extensive analysis of your solution. Then, with the information from the MAP Worksheet (found in Chapter 1) and SWOT Analysis Worksheet (found in Chapter 2), questions can be customized regarding the specific challenges of a targeted buyer. The remainder of the questions should reveal stated wants that relate to your solution.

When planning effective questions to ask the buyer, it is best to start with the outcome in mind. That being the case, the worksheet starts with problem-solving questions.

Problem-solving Questions:

- What "want" must a buyer state to lead to your solution as an offering?
- What future challenges might the buyer face that would lead to other features of your solution that have not been stated now?
- What other problem areas could we address?

Association Questions:

Explanation: Association questions analyze the problems that would lead the buyer to recognize the implications of the problems that exist, to become aware of his company's limitations in resolving those consequences, and to identify possible implications that future growth presents. These questions are designed to "create value" from the buyer's perspective. Because one or more of these six areas will have the highest value for the buyer, they also offer the most opportunity for your question development.

Write your questions below:

Gathering Questions:

Explanation: Gathering questions ascertain the information that you need concerning the stockholder's position in the organization, the person's role in decision-making, and behavioral observations about him or her. Keep in mind the initial cost to the buyer of your solution, as well as the long-term investment required. When formulating questions, make sure you understand the financial position of the company so you will know where you can add value that offsets expenditures.

The basic difference between investment and expenditure should be reflected in your questions: Your solution represents an investment if it returns value over time. These questions then revolve around the decision process, seller environment, any-time expectations, and initial recognition of present problems or future challenges.

This exercise has the most value when done as a team, but can be done alone. Either way, it is most valuable if completed before any calls are made. After you investigate the buyer organization, you will customize the questions for those stakeholders from your understanding of their positions and perspectives. That can be an individual exercise.

Write your questions below:

_____

_____

_____

_____

_____

_____

_____

Chapter Seven
Eliminating the Competition: How to Create a Winning Proposal

YOU'VE MAPPED the account, worked the politics, and demonstrated an understanding of the buyer's business. Along the way you have built bridges with key stakeholders by giving them the appropriate attention. They are interested in you—and in your solutions—and act as if they want to do business.

But that does not mean they will say, "We'll take it" and just hand you a check. It does mean you are now invited to respond to their Request for Proposal (RFP)—to put your solution in writing. The RFP defines the requirements of the project and its required solutions, plus the organization's terms and requirements for getting the business. Your sales attack now moves onto paper, as your team devotes its energies to writing a responding proposal, submitted usually to the Technical Evaluator.

Let's focus on Technical Evaluators and their job of vetting competing proposals. Doubtless they read them with rapt attention, one after the other, until they come across one that causes their jaw to drop in amazement. I can see them running down the street, clad only in a bath towel, yelling, "Eureka! I've found the solution!"

Okay, in the real world it doesn't happen quite that way. But it may be useful for you, the seller, to envision it happening for you, cheesy as it might seem. Because, contrary to popular belief, the trick is not to write a proposal that will win by being selected. Instead, you need to write a proposal that can't lose, one that the Evaluator will not be able to find any grounds to eliminate from contention.

Technical Evaluators spend their time eliminating proposals that don't work. They are trained to say "no" (it comes from running interference for the Financial Stakeholder), so if the Evaluator finds a way to eliminate all but one proposal, his job is done. Given that, your focus should be on writing a proposal that cannot conveniently be put aside.

Things to do include the following twelve items:

1. Spell out the benefits of your approach so the buyer cannot help but link them to the SECRET.
2. Having done that, supply facts to back up your claims.
3. Avoid long-winded sentences and paragraphs. Instead, use simple, easy-to-understand language. Don't write to be understood; write so that you cannot be misunderstood.
4. Completely fulfill all the RFP requirements, regardless of how ridiculous or insignificant some of them may seem.
5. Give the buyer what he wants in the form and format of his choosing, not in the way you think he should want it.
6. Let your strengths speak for themselves. Don't brag in the proposal about how great you are; let your Champion Stakeholder do that.
7. Know your competition inside and out, and then word the proposal so as to offset the buyer's strengths and exploit its weaknesses.
8. Admit that the buyer may consider you weak in some areas, and then do what needs to be done to counter those weaknesses.
9. Your proposal will sell for you when you aren't there. Give the

Evaluators ammunition to defend their decision (to favor you) to management.

10. Strategic sales proposals to large accounts are always buyer-focused. Tailor each proposal to the specific project and buyer.

11. Present buyer-specific applications, solutions, and benefits to define exactly how your products or services add value.

12. Ensure the proposal is easy for the buyer to read and follow.

Closing the GAP in a Proposal

There are sales teams that consistently outperform the competition. The "best practice" teams integrate the SECRET topics (covered in Chapter 5) to establish value for the buyer, based on concerns that were uncovered in the GAP questioning process. During GAP questioning the seller analyzes a prospect's situation and provides the buyer with the information needed to make a buying decision—the same information that is needed to write a strategic, winning sales proposal.

When sales professionals combine GAP questioning with proposal writing, interesting changes result. For example, if you learn that the proposal must include a sophisticated cost-benefit analysis, you should gather appropriate cost analysis variables during the GAP questioning phase of the sales process. Then the proposal will actually define specific analysis requirements for the selling process.

Remember, the proposal does not dictate how you should sell the product. It just focuses the information gathered during the GAP questioning process and puts it in an easy-to-understand format for the buyer.

Is the RFP Wired for the Competition?

There may be times when you will get the impression the RFP has been written in such a way that its terms favor a specific seller (other than yourself). In other words, it's been "wired." This happens less often than some people would lead you to believe; it is, after all, a handy excuse for losing. But it does happen, and you may have grounds for suspicion if the following trends show up in the RFP:

- The deadline to respond to the RFP or the rollout schedule for delivery seems unrealistic, either too stretched out or too aggressive. It could be that another vendor has convinced the Evaluator that the schedule given by the RFP is actually realistic, and that Evaluator is in for some trouble down the road.
- The RFP contains detailed specifications that seem to favor one approach to the problem, conveniently used by a specific vendor whose sales representative has been whispering in the Evaluator's ear.
- The RFP contains a surprisingly specific set of personnel requirements. Sometimes an RFP will require bidders to propose a project team that will include certain specializations. Oddly enough, these amount to perfect descriptions of people on the competitor's sales team.
- The RFP's estimate of the level of effort in man hours/days, or "man loading," required by the project may seem disproportionate to you. If it's higher than you think it should be, perhaps someone has convinced the Evaluator that more hours, and hence more money, are required. If it's lower than you think is reasonable, someone might have convinced the Evaluator that they could do the work cheaply.

As you go along, you will learn how to spot details that look suspicious. If you decide an RFP is truly wired and the vendor is already selected, then you should not propose at all. That said, very few RFPs are truly wired. Most companies holding competitive procurements are serious about keeping them competitive.

How to Wire the RFP

No one, however, ever said you should not try to wire the RFP to favor your own organization. I'm not suggesting you subvert the buyer's competitive procurement procedure. I am suggesting to be in there influencing the Evaluator, giving ideas, listening, asking questions, shaping the RFP's requirements, and proposing solutions. It's great when an RFP hits the street and you see some of your words in his request for proposal. Job well done!

The truth is that some IT projects based on "competitive" proposals are really won before the proposal is written. When they put out the RFP, the Evaluators are just going through the motions. The question is: How do you get into the position where the RFP is written for and about your solution?

One answer is to prepare a preproposal "White Paper Report." This report is written when you first get wind of a project, the earlier the better. Basic elements of a White Paper Report should include:

Stating the Problem

Discuss the problem in writing. Using the responses to your GAP questioning and linking the buyer's wants to the SECRET values, you have figured out what the buyer wants to have done. At this point be careful not to misrepresent yourself, but if you are using the Evaluator's words make sure that you are doing so in a way that is not patronizing. "He thinks the problem is X, when in fact the problem is Y," is not the best approach.

Simply state the problem as best you can from the buyer's point of view. A clear representation of the problem will establish you as a listener and a consultant, rather than a know-it-all expert. Equally important, it establishes your understanding of their stated wants and gives you the opportunity to tie these wants to the SECRET value topics.

Identifying the Issues

In the next few paragraphs, identify the critical issues that must be addressed in order to solve the problem. Every solution has potential "deal breakers." These may be technical, and the solution may require a level of technical understanding that only experts, such as your team members, understand. Or they may involve landmines, such as unrealistic deadlines or political infighting, that lurk just beneath the surface. Whatever they are, identify them and explain how they will be overcome. Use real examples of other implementations that had a similar impact and effect. By identifying stated wants and then explaining how to resolve them with your solution, you show the depth of your experience. At the same time, you educate the buyer. After reading your proposal, the Evaluators should feel they have actually learned something about their own project.

Solving the Problem

Once you have identified the problems and have explained how they are resolved, suggest a project approach that uses your solutions. This is really nothing more than a "letter of intent" or a "scope of work." Here you should discuss the tasks involved in the project, devoting no more than a paragraph or two for each. If you wish, include typical deliverables of each task and how long each task usually takes. But don't get into pricing, as this is not a proposal but merely an educational "White Paper" that the Evaluator can use as a reference when he writes the RFP.

Suggesting a Timeline

Finally, include an estimate of how long the project should take. It is especially important to include a timeline if scheduling is one of your critical issues. How long should each task take to complete? Which tasks are absolutely necessary for completion of the project? In what order should they be completed? This is a great opportunity to remind the stakeholders of their stated wants and how experienced you are at solving such problems and implementing these timelines.

White Papers are short; the best are no more than two to three pages. Try to adopt a helpful, suggestive tone, one that emphasizes your consultative approach. Who knows, maybe the next RFP you read will be wired for you.

Service Proposals

Instead of selling a discrete product, you may be selling an intangible service, such as hours of time. What such a proposal should focus on is whatever the prospect states he wants during the GAP questioning phase. Of course, things are never that simple, so here are some guidelines for putting together a service-related proposal.

How Will the Work Be Done?

If the Evaluators have released an RFP, they want to see how you will address each of the elements of the project. You need to go beyond just describing what you will do, however; explain how you will do it and why you have chosen to do it in a particular way. Then, make perfectly clear how doing the work your way will benefit the buyer.

Where Will the Work Be Done?

If you have specialized equipment or facilities that will make it easier to get the work done, stress them. On the other hand, if you have specialized equipment or facilities that are not appropriate to the job, don't dwell on them—you're just showing off, and no one likes to see that.

How Will It Be Managed?

The Evaluators will want to know how you will manage the work. After all, they want to feel the project is in the right hands and that you will handle their resources effectively. The management section is one of the most important in a services proposal. Most sellers pay little attention to this detail, and it costs them dearly in the long run.

Who Are the People in the Proposal?

Knowing the identity of the people who will do the work is important to most Evaluators. They want to be assured that you are bringing aboard people who have the right skills to provide the right service.

Dealing with a Short Turnaround Time

You have probably been involved in situations where the time available to write the proposal is ridiculously short. Perhaps the buyer did not give you much lead time, or maybe the RFP sat on someone's desk too long before a decision was made to proceed to the bidding stage. Regardless, the trick is to make the most of the time you do have. One way to do that is to treat the proposal effort as a project in and of itself. Consider adopting one or more of the following initiatives:

Prepare a Proposal Project Schedule

When people see a timeline with specific names and due dates, the project becomes more "real." Include the following milestones:

Two Drafts. Two will get you by. Three drafts are better, if you have time. Deadlines at least make you rein in the drafting process; the seventh draft is usually about the same as the third draft.

Team Reviews. You need to review the drafts. Each time you review a draft, the proposal will improve. And have a good proofreader go over each draft. Spelling errors are not acceptable; in fact, they can kill a deal.

Design Graphics. This can take a lot of time. Therefore, don't do what most teams do and put off graphics until the end as a "finishing touch." Remember, a picture is worth a thousand words. Graphics can help you get across a complex set of ideas in a very compressed space. Evaluators like graphics that help them see the points you are trying to make, instead of making them work hard to get the point on their own.

In the sales arena, the bottom line is that people buy based on visual stimuli, so using graphics can be your way of making the proposal stand out. Coordinate dates for drafts of the drawings, charts, and photos you plan to use in the proposal.

Administrative Support. This includes final word processing, copying, and binding.

Identify the Participants

The entire sales team should get involved, including the proposal leader, executive staff, graphics designer, and technical and subject matter experts. Also identify any administrative support you will need to put it together. Do you have subcontractors working with you to provide products or services to complete your solution? If so, should they help write the proposal? As early as possible get those involved in the process; it will increase the chance they will be there when you need them.

Hold a Kick-off Meeting

A kick-off meeting can serve to organize the selling team and set the pace. Writing a proposal is a project, and the proposal leader should be in charge.

The proposal leader should call an internal meeting to discuss the roles, responsibilities, and timelines of the project. Make sure to cover the strategic approach, giving assignments and follow-up deadlines. During the course of the meeting, identify the buyer's concerns and how you will resolve them. Determine the likely competitors and what you think they will offer using the SWOT analysis, and begin mapping out the clear and distinct benefits you offer the buyer using the

SECRET values worksheet in Chapter 5.

For short proposals, meet with your team for an hour or so to figure out who needs to be involved and what the timelines should be. More complicated proposals require more planning, but the extra effort will pay off in the end.

You wouldn't run a technical project without a plan. And your proposal projects are too important to your company to conduct them without one.

Make Go/No-Go Decisions

Should you go for it or let the competition have it? There will be times when you see that responding to an RFP will not be worth the effort. Always leave yourself the option of pulling the plug on a proposal project. But you should be making clear-sighted decisions based on what you know about the buyer and the competition, rather than on gut instinct or misplaced optimism.

What to Do When the Heat Is On

Sometimes the environment can get extremely competitive. You are at the final stages of a proposal, three companies are vying for the same account, and one of them is yours. How do you win the deal? What can you do to give yourself a competitive edge in the final phases of the sales process? There is no magic, but try the following tactics:

Respond Quickly

Evaluators assume (not unreasonably) that the salesperson who responds quickly is part of an organization that will also respond quickly when the time comes to support the product. It is always best to strike decisively. In many sales situations, the urge to buy fades when the Evaluator moves on to other challenges. New demands on the Evaluator's budget can soak up the money that had been allocated for your sale, or the Evaluator can forget why he was so interested in your solution in the first place.

Have "Price" and "Profit" Items

Any large project will include items that are both so popular and price-sensitive in that field that they are commonly sold at or below cost. These are "price items." Price items include some hardware products, training, or presales consulting. Specific products or services that you count on for margin are "profit items." Examples include software and professional services.

If your proposal is uncompetitive in terms of price items, your quote might not even be considered, no matter how reasonable the total deal is. On the other hand, when selling to a new account, great deals on price items can get you in the door. If you only sell price items, your company will go bankrupt, so you have to know which products to price aggressively and which to price for profit.

Sell the Benefits of "Profit Items"

Price items are often sold at the expense of profits. When you hear a seller say, "I can get you the necessary Ethernet cables at below wholesale" or "I can find six Oracle developers at half the cost," you know those are price items. Your competitors will probably be focusing their selling firepower on price items, since Evaluators are familiar with them. But price items will not pay your commission over the long term. And buyers, especially Financial Stakeholders, will not see your value in the sale if you talk about price items too early.

Profit items, on the other hand, are the solution-oriented seller's dream. Here is where you the seller can create real and significant value in the sale by promoting the profit items that enhance your solution and give benefit to the buyer. Do this by relating the profit items to what you learned through the SECRET topics and the GAP questions, covered in Chapters 5 and 6, respectively. Selling the benefits of profit items should sound like, "It will save you time by increasing throughput by 5.5 percent," or "It can save you thousands of dollars in future liability by adding a needed layer of security."

Offer Three Solutions

When assembling the proposal, consider offering more than one solution; three would be ideal. Most salespeople today sell a "system" of products that work together to provide the best solution. Why not offer several solutions in two or three quotes? (Evaluators will usually go for the one in the middle.)

Standardize Your Disclaimers

A proper disclaimer on a quote tells Evaluators that you are thoughtful. It can also increase your sales and save plenty of grief in the future. The problem is that IT buyers have a habit of saying things like, "It's your fault that I lost all my data because you didn't tell me to buy a tape backup/anti-virus package/surge suppresser/whatever." A standard disclaimer on every quote can note, "For your protection, we recommend the following . . ." The disclaimer should include things that are optional but may be necessary should reliability become an issue.

Your efforts may actually succeed, so give your Evaluator a chance to say "yes" by including an "Acceptance Box" with your fax number at the end of the quotation: "I accept the terms of this quotation." This is a written way of asking for the sale.

E-mail Your Quote

Vendors are finding that Evaluators respond best to e-mail. It's faster and more efficient than most other forms of communication. The standard format for documents transmitted electronically is Adobe's Acrobat Reader, a free software program that your Evaluators likely already own or can download from *www.adobe.com*. Attach your quotes to e-mail as an Acrobat document. The software needed to put your proposal documents into Adobe Acrobat format is not available for free at this time, so make sure your organization has acquired it before crunch time.

Follow Up with the Buyer

The best Evaluators are always impressed by thorough follow-up. That's why I am always surprised by how many salespeople say, "She knows what she wants. I don't want to bother her." Maybe it is fear of rejection that makes a seller say that. Or rank stupidity. The truth is, the slightest miscommunication can derail your sale. Checking with the buyer can get things on track again. Failing to follow up can cost you the chance to get "mind share" with Evaluators.

Follow Up with the Pursuit Team

Everyone agrees it is important to have a pursuit team meeting after a proposal to learn what was done right and what was done wrong. Except that hardly anyone gets around to actually calling one. "No time right now," they say. "Too busy." "We'll get to it later." And they don't. Guess what? They go out and make the same mistakes on the next proposal.

Organizing this meeting is difficult; I have to kick myself in gear sometimes to make room for it in my own schedule. But a postmortem is a key element in a proposal project for one simple reason: It helps you write better ones in the future.

Take one hour, at the most, as soon as possible after the completion of the proposal. Look at the proposal itself and examine how it was laid out, how well it sold its message, and how easy it was to read. Analyze the proposal preparation process you went through; identify what worked well and what needs improvement.

Not everyone involved with the proposal needs to attend the session, but the Proposal Leader must be there, plus anyone else who played a major role in preparing it, including the person who wrote the management section, key subcontractors who you intend to use in the future, and the graphic designer.

Topics to cover can include:

- Did the proposal look attractive and professional? Did it reflect the image you want your company to convey to your buyers?
- Did the proposal present a clear, consistent, and integrated plan covering the technical, management, and cost considerations?
- Did it successfully sell your ideas?
- Was the proposal easy to read? Was the writing style consistent throughout? Did you use graphics effectively?
- What could you do better next time?

These are not all the possible questions, nor does every question apply to every proposal. The point is, you should be asking questions such as these after the proposal leaves your office. Doing so will let you make significant changes and improvements to your proposal writing process. If you don't feel like holding postmortems, just think: Your competition is probably holding them and getting stronger in the process.

The Winning Proposal

Writing a winning proposal is an integral part of the sales process. Remember your goal: to produce a proposal that cannot be eliminated. To achieve that, the entire pursuit team must understand how the solution addresses the buyer's stated wants, and offer the Evaluator no possibility of putting the proposal in the reject pile.

Like anything, good proposal writing will result from practice, which leads to the final lesson in this chapter: The more often you quote, the more you will sell. Too many salespeople fixate on one deal, and that prevents them from working on more deals at the same time. While it is important to maximize your effectiveness, there are many factors involved in a sale that you just cannot influence. Regardless, keep the proposal team busy!

Resource File: Sample Proposal

Rather than write about how a proposal should be composed, I will present a framework for a proposal, including section headers and notes about what should be contained in each section. This is an idealized example; take it merely as a starting point or framework for any proposal you write. Buyers may provide a specific format to follow but, in case they don't, include these elements. After you get into your own project, it will become clear what sections need to be expanded, deleted, added, or otherwise modified.

Proposal Design

A key to a successful sales proposal is to keep the reader reading. One way to achieve this is to make your proposal easy to get through. This should drive your writing style. Try to achieve a "clear" style. Most of us know clear writing when we read it, though it may be hard to define. Clear writing is:

- Easy to read
- Efficient
- Easy to understand

Clear writing makes it easy on the readers; they can see what you're getting at quickly and they do not have to waste time trying to figure it out. This will make them happier. And that's what you want. Given the complexity of some proposals, you should discuss the details of your response with product management staff and other pre- and post-sales technical team members. You will find valuable written materials that can be used as a boilerplate for your proposal. They can also be helpful in the creative writing needed for your response.

Dress up the proposal by using photographs and graphics throughout: People buy visually and photographs will stir the imagination of the buyer. Also, pages get separated easily in large proposals, so guard against this by putting your telephone number and e-mail address on each page of the proposal, making it easy for the buyer to reach you. Do this even if you are e-mailing the proposal, since it will likely be printed.

Make sure your proposal is professionally bound and the appropriate amounts of copies are printed for distribution. Use the buyer's corporate colors as often as possible.

Cover Sheet

A Proposal to Enhance the Information Technology Effectiveness of XYZ Corporation

Scan in the buyer's corporate logo and insert it over the name. Have a running header or footer on each page with the proposal leader's name and phone number, in case the pages get separated. Note that the title emphasizes a solution benefit, rather than the product itself. It does not say "Proposal for Product ABC for XYZ Corporation" or "Proposal to Sell Information Technology Equipment to XYZ Corporation."

Table of Contents

Use the second page for a table of contents. Your word processing software may be able to generate one simply by cataloging the section headers. Then begin the third page with the Situation Overview, which follows.

Situation Overview

Begin the proposal by directly addressing the stated wants of the Financial Stakeholder. Couch the prose in second person, with the Financial Stakeholder as the target reader. During the gathering phase you should have made note of key phrases used by the Financial Stakeholder; slip them into the text, with emphasis if that seems appropriate. Using bullet points can enhance readability and is very eye-catching for a Financial Stakeholder. Points made about the desired outcome of the solution should also be based on stated wants that were uncovered in various sales meetings, but some will always be unstated wants that are inherent in the situation—the overall solution requires them. The section might sound, in part, something like this:

> *The XYZ Corporation is embarking on an initiative to further improve the performance and reliability of its back-office systems. During our discussions, you expressed the need for high-speed, modern LAN technology to serve as the basis for a corporate network, now and for years into the future.*

Below are observations of this situation:

- XYZ Corp. seeks to develop a leadership position in its niche.
- A key factor in this initiative will be the ability to provide customers with online order entry.
- Online order entry offers no benefit to the corporation unless it is integrated with back-office systems.
- The platform of the existing system does not permit this.
- Et cetera, et cetera, et cetera.

Through implementation of a solution that enhances the back-office network, we anticipate:

- Throughput and storage sufficient to support modern ERP and accounting systems.
- Easy interface to Web e-commerce functions.
- Et cetera, et cetera, et cetera.

We plan to provide a truly value-added solution for XYZ by providing the latest technology, et cetera, et cetera, et cetera.

Who Are We?

At this point you'll need to put in the usual prose about what your firm is and what you do. It may be boilerplate but that does not mean that it can't be interesting, lively, and informative. If your firm has sales history with the proposed solution in the prospect's business niche, stress that up front. However, do not name reference accounts here; leave that for the last section.

Successful, reliable implementations for businesses wanting long life cycles for their technology investment has been the hallmark of The John Doe Group, which has been serving the industrial base of Upper Metropolis for more than eight years. . . .

If staffing is an issue, stress your personnel and their experience. If the technology is not mature, stress your role as a pioneer. If the technology is mature, stress how widely accepted it is, and your experience with it.

If you are proposing complementary products from other vendors, you will need to have a separate section on each.

Solution Fundamentals

Having introduced the players, you now lay out your approach to satisfying the stated wants expressed earlier. State in broad non-technical terms (or define your terms as you go along) what is required to achieve the goals stated in the Situation Overview section. Each technology involved might be given a subsection. Be sure to mention the prospect's name.

> *Achieving the goals of XYZ Corp. will require removing its twenty-year-old 2.5-megabit network (which has been found to be incompatible with any mass-produced hardware made after 1996) and replacing it with modern technology. With 85 percent of technical support staff's time spent on crisis-response maintenance, further investment in the legacy system must be considered counter-productive. The building will have to be rewired. . . .*

Recommendations

Here you spell out your recommendations, but in terms of deliverables and benefits, rather than in operational details.

Increased Reliability
- Install a 100-megabit LAN with the latest routers, with enough servers to handle 25,000 transactions per day.

E-Commerce Enablement
- Install and configure a broadband Internet gateway as part of the network.

Et Cetera
- Et cetera, et cetera, et cetera.

Implementation Strategy

Now you approach the details of the actual implementation. But before you lay them out, you explain the groundwork involved in the implementation.

> *It would be simplest (and the results more aesthetic) to tear out the old system and install the new one. But that would result in unacceptable downtime for the entirety of XYZ Corp. Therefore, the new system must be configured and tested in parallel with the legacy system. However, since the building's wiring closets and cable conduits are already full with the legacy system's hardware (including vast numbers of undocumented modifications), installation cannot be finalized until the old system is removed. Therefore, some temporary inconvenience for the staff is unavoidable. . . .*

Implementation Through Phased Installation

Now we get to the details of implementation. Start with an overview and then show the actual steps. (If there are many variables involved, you might conclude with a description of what you would consider an ideal implementation.)

> *The hardware of the solution we propose actually takes up considerably less space than the legacy system it will replace. Therefore, a parallel installation until switchover is entirely practical, especially when using a phased approach.*
>
> ***Phase One*** *(two weeks): Install the servers on tables next to the wiring cabinets and run the cables down the edges of the aisles, taped for safety.*

Phase Two *(one week): Test the configuration.*
Phase Three *(six weeks): Migrate the applica-*
tions to the new servers one at a time.
Phase Four: *Et cetera . . .*

Return on Investment (ROI)

Here you give details on what the buyer can expect for the dollar investment over time. See the Web site *www.ejustifyit.com* for help in making these calculations.

Pricing Considerations

Here you list the cost of the hardware, software, and services that you proposed. Refer to the amounts as "investments," not "costs." Do not give a total since that will depend on decisions made by the buyer. Do, however, give accurate and complete pricing information, so the buyer can confidently calculate a final price.

Our Clients Include . . .

List your clients, hopefully recognizable names in the same industry as the prospect, or users of the same business or industrial processes. If you are selling for a start-up with no sales history, emphasize the experience of the principals.

Chapter Eight
Engaging the Enemy: How to Beat the Competition

SALESPEOPLE ARE THE MOST COMPETITIVE CREATURES on Earth: They hate to lose. The best ones identify a rival and will use every resource to shut them out of a deal. But before you, too, start charging out and launching full-scale warfare, "know thy enemy."

If there is a market for your solution, there is always competition. Even start-ups in new and unproven markets have enemies, competitors who are out there right now, finding buyers and making them happy. And that's a good thing, since without competition there is no market. (The worst thing that can happen is to hear a teammate say, "We don't have any competition.") Find out who your competitors are, what they are doing, and what you can do that is better—before they do the same to you.

In today's market, with the intensity of competition, it can be almost as important to stay current with your competitors' activities as it is to keep track of your own. A working knowledge of their strengths and shortcomings helps increase your confidence, not to mention your ability to devise strategies that get results. Once you know your rivals' position in the marketplace, you can take more effective measures to improve your own. By analyzing competitive moves, companies can anticipate market developments rather than merely react to them.

The "Cost" of Competitive Information

Information is the raw material of good decisions. Profits result from making good decisions, avoiding mistakes, and minimizing risk. The cost of obtaining information, in time or money, must be appropriate to the purpose. A decision requiring significant outlay of resources, such as entering a new market, targeting a new buyer base, or exploring an acquisition or merger, requires in-depth investigation. On the other hand, the background needed for an upcoming first meeting with a prospective buyer in an industry that is unfamiliar would probably require a minimal amount of information, sufficient to give you a basic understanding of the landscape and let you ask appropriate questions.

Yes, information costs. But when properly used, an investment in competitive intelligence can pay off quickly. For example, Compaq made a huge investment for its competitive intelligence (CI) program, allowing the company to quickly assemble a vast electronic library of analyst reports and competitive data. Employees can access about 80 percent of the resulting information on Compaq's own intranet, while the 20 percent intended only for the senior leadership team is kept on a separate server.

Small businesses can outsource CI to boutique consulting shops who specialize in analyzing competition and ultimately helping sellers make go-to-market sales strategies. Every successful IT-selling business, large and small, spends money collecting competitive information. Does yours?

Technological Versus Bricks-and-Mortar Competitors

Anticipating competition requires you to change your definition of "competitors." You can no longer afford to focus only on the players in your industry, providing products and services similar to yours, and prospecting the same types of buyers. Sellers today expand the definition of "competitor" to include any company that is capable of meeting your buyers' needs as well as you can, or better.

Technologically adept competitors searching for new markets to exploit have radically changed the competitive arena. Their ability to change, at a moment's notice, puts us all in a much different world. Nothing is sacred. The old saying, "It's a jungle out there," has never been truer. It is chaos; no one is safe and nothing is protected.

Most of us are familiar with bricks-and-mortar competition but not with technological competition. Mainstream competition usually takes the form of a couple of competitors battling in a predictable way for their share of an established market: Lilly versus Pfizer, WorldCom versus Sprint, American Express versus VISA. Compared with the highly unsettling technology competition, mainstream competition seems tame and, in fact, downright civilized. These mature competitors typically have been in business for some time and understand one another, having faced off competitively many times before. Because of this, they tend to act slowly and methodically in competitive situations, and keep score by analyzing "market share" and "production technologies" on a macro scale.

By contrast, technological competitors consciously use technology as their primary competitive weapon. They apply technology to create new markets and to gain access to widely different markets. One aim of these companies is to "totally replace" existing competition using leveraged, technology-based "efficiencies of scale." These competitors are ruthless and are never satisfied with their existing share of the market, their profitability, or the capabilities of their products or services. They come from the least expected directions to attack, and breach defenses that were previously considered unassailable.

The most successful ones create partnerships with their own

competitors in an effort to grow whatever industry they serve. They also use disciplined financial and strategic analysis to determine how to enter a new market and when to exit an existing market.

Competition can emerge from anywhere, anytime. Here are some examples. Internet-based services such as Headhunter.net and Monster.com have completely changed the recruiting industry. These services compete directly and successfully with the help wanted advertising in local newspapers. But the Internet services cannot rest, because the large newspapers are, in turn, placing copies of their own help-wanted ads on the Web.

Amazon.com has led with first-market mover advantage against major bookstore chains with its online book reviewing and ordering system. Yet the chains had only recently overwhelmed the independent bookstores by bringing to bear superior inventory and ordering technology. Amazon.com, meanwhile, has begun competing with garden stores, hardware stores, drugstores, pet stores, and who-knows-what other retailers. "Getting Amazoned" has arisen as a shorthand term for being surprised and overwhelmed by an unexpected competitor.

AOL, beginning as an Internet Service Provider a few short years ago, rose to become an Internet juggernaut with more than 25 million subscribers worldwide. Then it gobbled up Time/Warner to become the largest content provider on the planet.

All of these unexpected competitors entered markets that were new to them. They offered products and services that were equally new, all made possible by technological advances. On the other side were strong, entrenched competitors who were surprised by how easily other firms could use technology to restructure the competitive environment. Technological competitors always seem to violate the "rules," and that is what makes them so dangerous.

What should be of more concern is that the traditional techniques used for competitive analysis cannot warn us in time that technological competitors are about to "invade our space." New tools must come along that will let sales professionals defend themselves.

How do you tell if you are under attack? Better yet, how can you tell if you will soon be under attack? Perhaps your company is in the crosshairs of a technological competitor right now. If the industry in which you operate is ruled by mainstream competition, it is a candidate for a technological competitor. These new competitors can quickly grab high-margin niches by winning buyers with new, high-value products. Look around: Scanning the horizon to find them must be a regular, active job. The first line of defense is full alertness and anticipation. The best technological competitors, such as Microsoft and IBM, tend to play in many markets. If technological competitors have begun to target your markets, they must believe that they have something your buyers will prefer.

Technological Strategies

Technological competitors are willing to use strategies that are completely different from those that have been relied on by their mainstream competitors. They often compete in ways that put you at an upside-down disadvantage. If you sell based on low price, they will sell on high performance. If you sell a stand-alone product, they will sell a system. If you sell a service, they will sell reliability that requires no service. If you sell through distributors, they will sell direct.

Another strategy used by technological competitors is that they make moves that neutralize the investments of their new target competitors. If you sell powerful e-commerce software, they will offer a solution that eliminates the need for software. If you have invested in training distributors' sales staff, they will put the information on a compact disc and give it away. They will try to turn your investment into your Achilles heel.

Technological competitors try to offer products that completely change the way things are done—in their favor. If you perform a service, they might design an equipment module that fits right into the product's manufacturing line and performs your service automatically, on the assembly line. Usually, a technological competitor will address buyer needs by using an incompatible technology that you and your friendly mainstream competitors cannot easily duplicate.

Watch out for an unfamiliar competitor that approaches your prospects with a new solution that eliminates any need for what you sell. Many times technological competitors satisfy some of your buyers' needs better than you do. They can eat your market in bite-sized chunks. If they are smaller players, they will want to avoid head-to-head battles, so they will nibble away at the niches. They are methodical and careful in where they compete because they take your competitive clout seriously.

Keep a list of proven technological competitors that you spot in adjacent niche markets. Ask yourself, "Could any of these firms serve any of my buyers with any product or service?" If you answer "yes," watch them like a hawk.

Ways to Spot a Technological Competitor

Most of the time technological competitors will have little in common with you, so they are difficult to spot before they surface in your niche. They attack from different directions and can be spotted most readily if they compete in the same field as you. Here are a few ways to spot these competitors from afar:

- They already sell a different product to your buyer that solves the same type of problem that you solve.
- They offer solutions similar to yours, perhaps using the same platform but deploying different applications.

- They distribute different products and services through the same channels.

Your buyers, vendors, and channel partners are the best sources of information for advance warning of new competitors. However you spot them, once spotted you need to take them seriously, since they can pose a huge threat. Do not make the common mistake of minimizing their importance or dismissing them as "insignificant" or "remote." Sometimes I hear sellers dismiss a possible competitor, since its products lack a specific feature or function, even though adding it would take a trivial amount of time and effort. Once ignored, new technological competitors can work for months or years before anyone takes them seriously, and suddenly a formerly disdained product has become a heavyweight competitive threat.

Listen to your buyers carefully, and take their advice seriously. Discuss with them any other companies already doing business with them, in any step of the sales process that could involve your product or service. Keep an especially sharp eye on your most innovative buyers and suppliers. Innovators are often the first firms approached by technological competitors looking to test a concept or a new application. Hearing about it can earn you advance notice of six to twelve months.

When you call on your competitors' buyers, ask for their objective assessment of your competition. Ask open-ended questions that will get the buyer talking about how your competitors approach buyers and prospects, their selling strategies and ideas, and the quality of the service they provide. Find out what that buyer particularly likes or dislikes about their current supplier. Listen attentively; the information you collect might help you devise a foolproof selling strategy for that buyer. What you know about a prospect's relationship with the current vendor can turn their buyers into yours.

Attend Trade Shows and Conferences

Conventions, trade shows, seminars, and other professional meetings give you the perfect opportunity to talk with your competitors. Get a list of the attendees and copies of presenters' papers. Take special note of atypical companies that get coverage in your industry's trade press. Watch industry participants doing anything out of character. Monitor any merger and acquisition activity that affects your industry. Large companies can easily buy a small company as a way into your market, or use the small firm's technology in new ways or new places to do you in. Many companies use acquisitions to gain access to unfamiliar markets. Cisco, for example, has been building a conglomerate of companies, rolling up technology leaders to capture markets, quickly surprising the competition and dominating the landscape.

In my first IT sales job, with Novell, regular attendance at the local networking events and venture capital meetings brought me face-to-face with competitors. My conversations with them yielded valuable insights into how they presented themselves and their products to our mutual prospects. Familiarize yourself with their literature, displays, and advertisements, and try to determine their selling ideology. Information on their company philosophies, or predictions on the future of the industry and their roles in it, might help you design marketing campaigns or presentations that differentiate your company.

Do not be afraid to share information on your own company and position as well; inside information from your competitors will be worth any "trade secrets" you give away.

Watch Your Smallest Buyers

Good technological competitors know that the buyers you would like to ignore are their easiest pickings. Their sales costs will be low because such marginal buyers will run to new products that look faster, better, or less expensive. These buyers represent a perfect spot for a new competitor to enter your market. So see who offers them something new, and anticipate technological competitors at those accounts.

Benchmark the Best Practice

Benchmarking is not industrial espionage. It is the art of finding out, in a perfectly legal way, how others do something better than you. The ultimate goal is to imitate, and perhaps improve upon, their techniques.

Benchmarking or "modeling" the competitor's best practice involves mirroring the moves of the competitor, establishing the features of the competition's products and their pricing model, and then improving the process. These days, the first part is easy, thanks to the Internet. Wade through the competitor's Web site and learn all that you can. You'll likely not find its pricing structure posted on the site. (That anyone ever does in the business-to-business competitive environment is amazing, since they, and you, should be selling their wares on the basis of business value, not price.) So call the company and dig out their pricing model.

Benchmarking may be the single most effective tool used by organizations that have improved performance. Major organizations, including such corporate giants as IBM, AT&T, Eastman Kodak, Motorola, and Xerox, have numerous benchmarking studies in progress right now. IBM, for example, has already performed hundreds of studies, most of them in the past few years. Small organizations can benchmark as well at a very low cost.

Mine the Web for Information

There is so much information on the Web; here are some useful ideas to cut through the clutter. For instance, look at The Wall Street Transcript (*www.twst.com*), a service that tapes senior management's presentations to analysts and makes them available for sale. You will frequently find senior management talking about goals and assumptions. Also, company Web sites will have an "about us" area full of executive biographies and information about mission, vision, and values. The pictures they draw may not be completely accurate—you have to expect favorable spin—but they offer a foundation to build on as you go further with your analysis.

Additionally, a number of Web companies offer services specifically designed to keep you aware of your competitor's goings-on. Most of these services are Web sites that you subscribe to. As a subscriber you can set up a profile of your interests. Intelligent software agents within the site will then comb the Internet and other sources to deliver you the latest information on those search terms. Each time you check in to the service you will receive an update on what has changed.

Look in the "Sources of Information" section in the Appendix for related Web sites.

Play the Spying Game

Sorry, you're not in the CIA. Besides being illegal and unethical, dirty tricks such as phone taps, dumpster diving, and executive surveillance are simply unnecessary in competitive situations. After all, it is estimated that 90 percent of the information you want is publicly available.

What about the other 10 percent? Your best source of fast, reliable intelligence is the competitors' salespeople. Because of their daily contact with buyers and a variety of other individuals, they remain your best bet for "one-stop shopping" to gather critical information. No other source is so consistently helpful in such vital areas as pricing, product innovations, sales practices, dealings with distributors, and general marketing plans. From there it is only a short step to gaining detailed insight into advertising and promotional plans, test market plans and results, assessments of markets and market share, sales and profit data, buyer lists, distribution schedules, and upcoming negotiations.

More Traditional Ways to Find Information

An excellent but frequently underused source of information on companies is the local newspaper from the company's headquarters or branch office. The executive's guard is often down when he talks to the hometown press, so better information on the company or industry may be found here than in larger publications.

Another overlooked source consists of lesser known trade publications and pricey subscription newsletters. The increasing interest in information, especially targeted information, sparks the launch of more than one thousand new periodicals every year. While you may not have the time to check out these other titles, they all contain substantive, valid information about the industry.

Finally, you may find worthwhile information in publications that are not known for their editorial content. For instance, CompUSA, a major chain of retail stores, publishes a monthly newsletter that is primarily an advertising circular. But its full-page article on buying habits of Generation X was substantive enough to be reprinted in an industry publication. What is the likelihood that you would read an article that appeared in a throwaway newsletter? And if you did, what is the likelihood that you would believe the content? Those who looked beyond their bias for "real" sources found a gem of information.

Sleeping with the Enemy

Right about now, you may be thinking, "Make love, not war." In fact, it is now not uncommon for sales managers to ask sellers to cooperate with, not kill, competitors. Why? Because buyers are demanding it. They need complex solutions that require solutions from multiple vendors, and they want a one-stop shopping experience.

Collaborating with competitors is common in the technology world, where buyers purchase multimillion-dollar enterprise systems that link together corporate functions, and companies in every industry are aligning themselves with their fiercest rivals. Given that, your channel partners or your buyers themselves could become your next competitors. Carefully watch the most innovative among your partners and buyers for any sign that they are offering something that competes with you. Monitor all the usual suspects. If the most innovative firms are public companies, read their annual reports.

If you are considering collaborating with another firm, the experience of a large Silicon Valley software manufacturer may be instructive. The company's sales team took a prospective buyer through every step of its brand new software program, which carried a $300,000 price tag. The buyer, in turn, presented the firm's proposal, line by line, to a competitor. Receiving the bid turned out to be only the door prize for this lucky competitor; the real jackpot was obtaining trade secrets that enabled the competitor to compete head-on in a market that the unsuspecting software firm dominated.

Ideally, everyone to whom proprietary information is presented should sign nondisclosure agreements. Confidentiality agreements should also be an important element in your business dealings with channel partners, who are often the most knowledgeable outside sources concerning your company. They know intimate details about your business's products, production processes, total outputs, and strategic plans, information that could end up in the hands of an existing or future competitor as these channel partners prospect for buyers themselves.

A key challenge to working with such unlikely allies is maintaining a relationship without divulging important competitive information. Pick your partners wisely, because information is your company's most valuable asset.

Summary

Detecting technological competitors early is no guarantee that you will be able to thwart their attacks. But early detection does give you the opportunity to seize the initiative, assess the threat with a cool head, and respond in a strategically appropriate way. It can mean the difference between preserving a strong position in your most-valued niches, and losing it all.

Of course, the best competitive technique of all for this new environment is to take the initiative and become a technological competitor yourself. Don't get headaches; give them, to your unsuspecting new competitors.

Resource File: Gauging Your Competitive Readiness

Before you make the expensive plunge to go to market, gather your team and discuss these key fifteen questions:

1. How is the competitor structured, and how do the pieces fit together?
2. Who are its key management players?
3. What is the corporate culture? What are its values?
4. What has it done in the past? What works?
5. What's changing?
6. What is the effect on our organization and products?
7. What is the effect on our resources (labor, technology, clinical advocates, sales/distribution channel)?
8. What is the effect on organizations and products with which we align?
9. What is the effect on organizations and products with which we compete?

10. What is the effect (if any) on organizations that are industry kingpins?
11. What is the effect of the external economic and social environments?
12. Does this affect our strategy or our tactics?
13. Does it change our plan or does it simply make us act differently?
14. What is the best way to coordinate our strategy and tactics?
15. Are we certain that the information that we are acting on is accurate? Is there a way to verify it?

Chapter Nine
Managing Your Buying Allies: From Rainmaker to Team Captain

RECENTLY I VISITED with senior managers of a Value-Added Reseller (VAR) to assist in developing strategies around its major account business. The moment I arrived, I was escorted to a large conference room crowded with sales and operations management interested in being involved in the discussion. Beginning the meeting, as is my practice, I inquired about the company's best-practice salespeople and about their largest buyers and how they are managed. After a few minutes I understood why the firm was looking for help—it had serious account management problems, making itself vulnerable to the competition.

The Aetna Life Insurance Company was the company's largest account, representing 18 percent of its sales, or $22 million in revenues. A manager named Mike handled the account. Mike was young and smart, and had clearly made or exceeded quota for two years running. By management's standards, he was a superstar. What concerned me was that all the information about the account was locked up inside his head. Nobody but Mike knew the critical details of the account, making it impossible for management to help Mike manage it. And Mike was treating this piece of business as if he did not want anyone playing in his sandbox.

The account had stagnated during the previous two quarters, causing concern among some of the managers, while others openly defended Mike's past performance. As we were discussing the account, I observed that Mike's account management style "made the company vulnerable to competition." A deathly chill fell over the room. Throats cleared, glances were exchanged, and feet shuffled nervously. After about thirty seconds of silence, which seemed like thirty minutes, the vice president of sales broke the ice. "I knew this wouldn't be pretty," he said. "But this is the kind of feedback we need." With that, the floodgates opened. There were four other executives at the meeting, and every one had strong opinions about the Aetna account.

"Mike has fallen asleep at the wheel," snapped a division manager.

"Fallen asleep?" Mike's direct manager replied. "He's the best we have! I'm supposed to consider him a problem?"

"Sure he's good," said another, "but he's too much off on his own. I trust him to sell our services with no questions asked. But he leaves too much up to chance and not enough audit trail."

The vice president looked up and glared at the managers. "You're right, I see Mike as a lone wolf," he barked. "I can't tell you how many nights I wake up thinking about our competitive vulnerability there." He paused before continuing. "This is it. The bottom line, I'm afraid, is that we just don't understand what's going on."

Do not make yourself vulnerable to this kind of situation. The fact is this sort of account isolation makes salespeople uneasy, too. After all, they could do so much more with proper support.

As it turned out, after management put pressure on Mike about the stagnation at the account, he decided to leave the company with very little notice, taking the valuable, undocumented account information with him. To compound the problem, management did not know anyone in the Aetna organization; they felt as isolated as Mike must have felt. The team building that we discussed in Chapter 2 simply had not taken place, and Mike had been able to hoard all the account information. That is not an uncommon attitude, but very few salespeople ever get as far as Mike had without the backing of a knowledgeable team and full management support.

As for Aetna, it decided not to suffer a potentially disruptive period of transition with an unknown new salesperson; it followed Mike to his new company. Seeing the loss of 18 percent of its revenue, the company's sales management team was hit with the realization that stagnation at this account was the least of its worries.

Adding Method to the Madness

The reason that communications broke down—and the buyer and the salesperson deserted together—is that this company, like many otherwise well-managed firms, lacked a consistent methodology, or process, for handling large account information. The company needed a systematic approach to those buyers, one that went way beyond spot-checking for obvious problems to include sales team planning and implementation.

Because it lacked such a process, the company was bound to run into trouble and sales were bound to stall. It wasn't that the management team in the conference room was consciously excluding Mike from their account management strategy, any more than Mike was consciously excluding them from his planning. The problem was more fundamental than that: No one had a strategy to begin with. Had they set out to devise one, there was no way to include everyone. My original suspicion had been correct: No one was really managing this account. Instead, the managers were getting by one day at a time, until the day they found themselves in deep trouble.

The task of managing important buyers is different from other sales programs. Important buyers—those that have a large effect on the seller's bottom line—require more attention from management plus a higher level of dedicated resources, so that solutions can be customized around that buyer's business.

As I discussed in Chapter 2, the salesperson should be part of a team that, through the entire sales process, continuously builds a relationship with the buyer. But once the initial sale is made, different members of that team should be part of an ongoing "partnering team." The buyer then has more than one contact with your organization. If one member of the "partnering" team spots an opportunity to grow the business or perceives additional buyer wants, that observation can then be shared and acted upon by the appropriate person. Again, however, someone should be designated "team leader."

In a smaller company, as is the case with many selling organizations, the team leader may continue to be the original salesperson. But wiser companies add at least one additional person to the team. That could be someone in management or in technical or consultative support.

"That Account Doesn't Need Management"

When a company thinks a large account doesn't need management, it's operating under the assumption that the account is fully under control and that the revenue stream from that account is guaranteed. That is the most dangerous attitude a seller can ever have. Pride and selling are two concepts that, when combined, are a recipe for disaster and will surely make loyal buyers ex-buyers.

The problem is not just "bad attitude." It's that, from the buyer's point of view, such confidence always translates into cockiness. If I think a current cash-cow account does not require my attention, I will inevitably communicate that feeling to them. And once they realize that the seller does not think that they have to work for their business, they start looking for other vendors. In today's competitive environment at our keyboard fingertips, alternatives will not be hard to find.

At the same time, some salespeople object to reporting and hands-on monitoring from management and other team members. They say that certain buyers do not really need ongoing management, stating that they are so solid they are "almost on auto pilot." But there's a world of lost revenue in that "almost." As I point out to my buyers, a nautical analogy would be to say that the sea in front of us is clear of danger and our ship is built so well it is unsinkable. As the fate of the Titanic showed us, that attitude could be reckless to the point of tragedy. Businesses, too, can hit an iceberg. Pay attention to your existing buyers, and in return they will reward you.

Be Present

By "present" I mean pay attention to the present. I stress this because most sellers who manage already existing buyers do not keep their eye on today. Many try to understand what is happening with their large buyers, but they fail because they are looking behind rather than ahead.

Take the typical "account plan" done for management by most salespeople. It is generally an accumulation of ideas wrapped in data intensive packaging, including such supposedly useful pieces of information as market analysis and the latest sales figures. The information is meant to provide the selling organization with an overview of what's happening in the account, but all too often the information doesn't do that. Why not? Because the information is old. At the start of the sales process, one of the main reasons to talk with the Financial Stakeholder and the Technical Evaluator is to find out what they see as their future goals. What do they expect the company to be doing in six months? One year? Five years? What business can you expect to do with them in the future? And when you sell solutions to a buyer, you are involving yourself in their future plans. With technology changes occurring at lightning speed, account history is just that, history. While we can learn from account history and the successes and failures that it reveals, managing buyers should be planned using the future, with regard to the history—not trying to repeat it.

I have seen plenty of thick account plans that turn out to be reviews rather than previews of what is happening. That's typical of account plans. Almost by definition they compile "instantly obsolete" historical data. Trying to bring control into a large account by relying on such dated information is like trying to invest in today's stock market by looking at the company's performance from last year. Even in those isolated situations where an account plan does address the future, the forecasts tend to come from past sales successes: "I sold them five hundred last year; this year I forecast six hundred." These forecasts are rarely accurate and usually misleading. That's not surprising, since they don't take into consideration today's and tomorrow's dynamic realities, including your buyer's wants. Therefore, they cannot help you manage your large buyers in an atmosphere of constant change at Internet speed. The only thing that can help you do that is a methodology that is itself dynamic.

Start in the present, not in the past, to predict tomorrow's new business.

Which Buyers Should You Manage?— Lessons from the Front

The Strategic Account Management Association (SAMA) recently sent a survey to 2,200 companies that sell to large mainstream, bricks-and-mortar businesses. About 220 replies were returned, representing a 10 percent response rate. Of the surveys returned, 194 firms were conducting strategic account programs to a degree necessary to be included in this study.

Vendors represented a wide range of industries, including banking/finance, business services, healthcare, manufacturing, telecommunications, software, and transportation. The company size varied, with sales revenues ranging from $3.1 million to more than $50 billion. The number of employees ranged from twenty-one to more than 150,000.

Table 9-1

What criteria are used to select strategic buyers?	
Volume of estimated potential business	92%
Volume of past sales	78%
Size of buyer	55%
Industry of buyer	49%
Volume of past sales with competitors	43%
Management discretion	29%
Buyer relationships with competitors	18%
Alignment for future growth	8%
Geographic location	5%

© Strategic Account Management Association

As you can see from **Table 9-1**, potential revenue turned out to be the key criterion for the selection of strategic buyers. In fact, analysis of the responses indicates that there are three patterns of account selection. We can say that companies normally choose strategic buyers based on:

- **Future potential**. These companies view a buyer in terms of the potential for revenue growth. Sellers use formal estimates produced by sales, marketing, management, or the buyers themselves to determine the potential business an account may generate. While formal estimates are important, the respondents indicated that management discretion and the size of the buyer are also critical for selecting a strategic account.
- **History**. Sellers choose strategic buyers based on their own his-

tory with the account or their knowledge of their competitors' history with that buyer.

- **Current business environment**. Sellers evaluate a potential buyer in terms of their current position in the marketplace. These sellers pay particular attention to the industry of a potential buyer and the relationships that buyer has with competition.

Table 9-2

How buyers were chosen
Estimates by sales management 66%
Estimates by buyers 48%
Aggregate forecasts from salespeople 41%
Analysis from marketing department 41%
Past volume of business with competitors 34%
Management discretion 19%
© Strategic Account Management Association

How Do Companies Actually Choose Strategic Buyers?

As **Table 9-2** illustrates, the criteria used to select potential strategic buyers is one thing, but actually choosing a strategic account is another. Apparently, making the right choice must be based on more than just management discretion (19 percent).

Formal sales estimates from sales management (66 percent), buyers (48 percent), salespeople (41 percent), and marketing personnel (41 percent) dictate the buyers companies choose to do business with. As the study indicates, estimates by sales management are the number one way sellers use to choose buyers. This confirms how important it is to provide accurate and substantive revenue estimates and account plans to sales management. Accuracy of these reports is critical to sales success.

Table 9-3

Who develops and manages strategic account programs?		
	Develops	**Manages**
Sales Executives	79%	55.1%
Mgr. Strategic Account Programs . . .	42.5%	34.5%
Account Executive	43.1%	5%
CEO	34.5%	4.6%

© Strategic Account Management Association

Who Develops and Manages
Strategic Account Programs?

Table 9-3 tells us that the sales executive has a key role in developing and managing strategic account programs. While the marketing head and the chief executive officer both are highly involved in the development of the strategic account program, once it is launched it is left to either the sales executive or the manager of strategic account programs to implement and manage.

Table 9-4

Role of strategic account manager
Team captain 76.4%
Rainmaker 54.6%
Coach. 52.3%
Strategist 32.8%
Consultant 2.8%

© Strategic Account Management Association

What Is the Role of a Strategic Account Manager?

The lesson from **Table 9-4** is that the role of the strategic account manager differs from that of a typical salesperson. Only about half (about 55 percent) of the respondents felt that the strategic account manager takes on the role of "rainmaker" or "prospector," that of stimulating the buyer's interest in the strategic account program; in other words, a typical up-front selling function.

An overwhelming majority (76 percent) of the respondents saw their strategic account manager as a "team captain," the person who marshals and coordinates company resources for the program. Another 52 percent felt one role of their strategic account manager was as a "coach," the person who advises and oversees the sales organization's execution of strategic account programs.

Respondents typically have their strategic account managers actively involved in the implementation of their programs; only 33 percent thought of their account managers as a "strategist" or someone who establishes general strategic account policies.

One of my buyers, a highly successful banking software company, employs members of its sales consulting team in implementation. As the consultants built a relationship with the front and back-office buyers, they became privy to new opportunities. Their largest account, Deutsch Bank, was managed in this way with continuous sales consulting support. As the implementation and consulting teams perceived new wants, the software was further developed to meet needs deeper into that account. This prevented the competition from getting a foothold and later, a designated person in customer support joined the team. The true account manager was in constant contact with all these team members as well as Deutsch Bank's teams. With this multi-person interface, the account is still solidly theirs, although the bank has had several personnel changes over the past few years.

Table 9-5

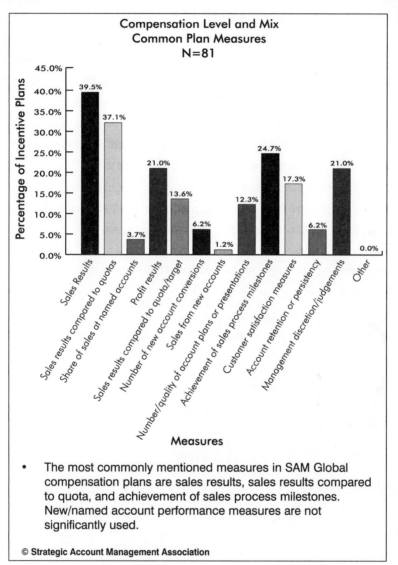

Compensation Level and Mix
Common Plan Measures
N=81

- The most commonly mentioned measures in SAM Global compensation plans are sales results, sales results compared to quota, and achievement of sales process milestones. New/named account performance measures are not significantly used.

How Are Strategic Account Managers Compensated?

We can see from **Table 9-5** that, for the most part, strategic account managers are compensated in the same fashion as most salespeople. Compensation based on sales revenue is a fairly common approach, with some strategic account managers indicating they are now paid on profitability. This table shows clear movement to compensation based on sales process milestones such as customer satisfaction and achievement of predetermined sales objectives. However, the sales quota is still going strong: Most stated that they carry an assigned sales quota, much like other salespeople. Half of the respondents said they have quotas larger than those carried by other salespeople.

Table 9-6

Do strategic account managers have formal authority over the rest of the sales organization?

Strategic account managers
  have no formal authority over
  others in the sales organization 47%
Sales teams report directly
  to strategic account manager 24%
Strategic account managers
  may assemble temporary or
  "virtual" sales teams, which
  report to them 22%

© Strategic Account Management Association

What Authority Do Strategic Account Managers Have?

The account manager has very little "formal authority" in the organization. Most account managers have a "matrix" authority having very informal influence with team members. **Table 9-6** shows that nearly half have no formal authority at all within their organizations, suggesting that their ability to sell ideas internally in their own organizations as well as solutions externally to the buyers are extremely important to the success of the management of buyers.

Table 9-7

How successful has your strategic account program been over the last year?
Very successful 19%
Successful. 53%
Neither successful nor unsuccessful 6%
Unsuccessful 9%
Very unsuccessful. 13%
© Strategic Account Management Association

How Successful Are Strategic Account Programs?

Considering the push from both the seller and the buyer for increased use of strategic account programs, it is surprising that **Table 9-7** indicates that only 53 percent of the vendors felt their programs were successful. Additionally, even though more sellers felt that their programs were successful rather than unsuccessful, only 19 percent felt that their programs were effective. This leaves a huge margin for improvement in the quality of strategic account programs.

Table 9-8

How do you measure the success of your program?	
Buyer volume	80%
Buyer satisfaction	53%
Profitability	45%
Volume of recurring sales streams	29%
Incremental orders from existing buyers	22%
Number of buyers with strategic account agreements	11%
Number of transactions/orders	6%
Number of units shipped	6%

© Strategic Account Management Association

How Is the Success of These Programs Measured?

As you can see from **Table 9-8**, most companies use some form of financial measure as the basis for determining program success or failure. When asked why they created these programs and how they chose which buyers would become part of the strategic account programs, they also cited financial reasons.

What Are the Organizational Barriers?

As **Table 9-9** indicates, sellers felt that the key organizational barrier to success was changing their buyer's past habits and opinions. While this is a difficult obstacle to overcome, the next two most popular responses may give some clues as to why companies are having such difficulty changing their buyer's opinions: forty-three percent felt that their resources are not restructured effectively and 41 percent thought there was resistance to change among personnel within their own organization.

Table 9-9

What organizational barriers does your program face?
Changing buyer's past habits and opinions . . . 44%
Not restructuring resources effectively 43%
Resistance to change by personnel in the organization 41%
Lack of long-term management commitment . . 24%
Being a late entrant relative to competitors 22%
© Strategic Account Management Association

What Are the Critical Success Factors?

Selling in the context of a strategic account relationship demands an unprecedented level of teamwork, the capacity to become intimately involved in the buyer's business, and a significantly higher level of business skills than other types of account relationships. While this study only uncovers some of the best practices and potential problems of current strategic account program management, it is clear that this environment is different from the traditional sales situation. Problem-solving skills are more important than traditional selling skills, and the ability to manage resources greatly outweighs the ability to close a deal.

While many members of sales teams need to exhibit more of the take-charge attributes discussed in Chapter 3, the account management team needs amiable personality styles to build the team concept both methodically and patiently. The team also needs analytical types, with their accuracy and dependability to support the technical aspects, and to make sure that everything works as planned.

Considering the importance of bringing resources from throughout the organization to bear on a single account, perhaps sales organizations should emphasize learning how to marshal the appropriate resources. Training that fosters a culture of cooperation, internal communication, and teamwork will also help companies deliver clear value to their buyers. When their buyers see and understand that value, revenue should soon follow.

Synergy between the company and the buyer appeared to be the key factor influencing the success or failure of a strategic account program. The SAMA survey defined synergy as a feeling of partnership, trust, and a philosophical alignment of goals and beliefs. The selling companies felt that the relationship must be a win/win; by meeting the wants of buyers, they could meet their own needs as well. On the other hand, the study shows that if a company defines success through financial measures, strategic account managers might engage in quota-driven, transactional sales behaviors that impede the building of long-term strategic relationships.

Table 9-10

Factors for a strategic account relationship
Successful
Management involvement 87%
Communication. 48%
Financial indicators. 31%
Buyer satisfaction 14%
Pricing 4%
Unsuccessful
Lack of management involvement 53.4%
Lack of consistent strategy. 36%
Financial indicators. 33%
Communication problems 24%
Lack of follow-through 21%
© Strategic Account Management Association

But as indicated in **Table 9-10**, when asked to state three factors that make for a successful strategic account relationship, an overwhelming majority (88 percent) stated that synergy ranked number one. Management involvement scored a close second, with 87 percent of the respondents citing resource and time commitment, as well as account intelligence and conflict resolution, as critical to success. The third success factor was communication (48 percent), which included open communication lines between both parties and active, consistent contact with key decision-makers within the strategic account.

Only 53 percent of the respondents used any nonfinancial measure of success. Additionally, most of the selling companies stated that their strategic account managers have an assigned sales quota and are rewarded by commissions paid on sales volume or profitability. Only 12 percent rewarded their strategic account managers based on the nonfinancial measure of buyer satisfaction.

While it is true that companies develop their programs to increase revenues and become more competitive, they may be losing sight of how important it is to bring value to the buyer. Before they can experience any revenue increases, they must build long-term relationships with their buyers. Therefore, perhaps one reason 55 percent of the respondents did not consider their programs successful is the discrepancy between their financial measures for success and the factors they see as essential for a successful strategic account relationship.

The way outcomes are measured tends to define sales behaviors. If a company defines success through financial measures, it is likely that the strategic account managers responsible for these buyers—as well as employees throughout the organization—are engaging in quota-driven, transactional sales behaviors that impede the building of long-term strategic relationships.

A Step-by-Step Approach

What are the next steps to implement a proper account management system? Here is a step-by-step approach based on the experience of some of my buyers.

Step 1: Chunk the Account

The first step in setting a strategy for a large account is to define a manageable field of play. I think of it as "chunking"—dividing the account into small bite-sized portions that are digestible and easily understood.

Most companies today grow by subdividing, merging, or acquiring other companies that can provide strategic value. Nike, for example, now sells not just its traditional running shoes but also sportswear, sports equipment, and audio and video products and is planning on introducing a more formal clothing line. Each company has a different management team and operating budget.

Consider, too, the case of one of my buyers, an enterprise software company providing services to the hotel and hospitality industries via the Internet. One of its most lucrative buyers is Marriott Corporation, which brought in nearly $10 million in revenue in one year alone.

Marriott is not just one company. It is a conglomerate of five operating companies each with separate managing budgets, including Marriott Hotels, Marriott Courtyard, Comfort Inn, Executive Suites, and Travel Services. Each of these operating companies is separate and distinct, and in some cases competes against the others. They each employ thousands of people worldwide and cater to different markets in different ways. No one seller can handle all of the different companies by themselves. Trying to do so would be confusing and would leave money on the sales table. So when my buyer targets "Marriott," it sets separate and distinct strategies, one for each operating company.

Step 2: Decide What to Sell Them

The second step is to decide what you should be offering this buyer in this niche. You cannot just hand them your catalog—the process has to be subtler than that. Taking that route implies you have limited interest in their real wants, and that you are there merely to take orders, from anyone, anytime.

Selling effectively to large buyers means reviewing what you've sold them in the past, distinguishing the good sales from the bad ones, and discovering which pieces of the solution have a genuine fit with the buyer's wants and needs. Sometimes it also means rewriting the "menu" you offer the buyer based on what you have discovered about the product fit and how it can meet unexpected or anticipated wants and needs.

Step 3: Partner with the Buyer

Once you've identified how your solutions fit your buyer's problems, you are ready for the final step: determining how you can "help" the buyer's business.

It is critical that you look at it in terms of the buyer's business and not your business. To effectively manage a large account, you must define "help" through the buyer's eyes and deliver value that enhances its bottom line. When you're selling what you believe to be a great solution, it is common for the seller to assess its value from the inside and not from the buyer's perspective. Value is whatever you can get for it. The buyer and their perception of what you can do for them justify that value. Looking at value from the buyer's view (remember the SECRET) is the only strategic way to help them—and you—succeed.

Summary

Managing buyers requires executive teamwork with the selling mentality and skills to "help" the buyer fill stated wants. This is done by looking at the buyer's business and mastering the team skills needed to organize the account. It also requires a significantly higher level of business skills than other types of traditional selling relationships. Account managers must be team captains, not traditional sales prospectors or rainmakers. For them, the ability to manage and leverage resources is more critical than being able to close a deal.

Resource File: Creating a Strategic Partnership

Here are questions to ask of your own company before embarking on a strategic partnership with a buyer:

- Does your company have the capacity to establish a separate strategic management team?
- If so, who on the selling team should also be included on the strategic management team?
- Do you have a technical consultant (or team) who will continue to work with the buyer after the initial sale?
- Is there a designated technical support person (or team) for your buyer?
- Have all these people been trained to recognize and uncover new wants and needs as they arise with the buyer's team?
- If your company does not have the capacity for a separate team, have you, as the salesperson, teamed up with someone from management to continue to uncover needs as your partnership with the buyer continues?
- Is there an established procedure to ensure that buyer information is accessible to someone other than you?

Chapter Ten
Coordinating Your Sales Channels: How to Get Leverage

UP TO NOW we have explored the IT selling process as a direct selling approach to the end user. Being able to use a direct sales process to sell to large accounts is of critical importance; large buyers demand it, but in pure form it constitutes only one "sales channel"—the "direct" sales force channel. Direct in that it moves a product or service from the creator or manufacturer directly to the end user. There are many other "indirect" channels, such as Value-Added Resellers (VAR), Original Equipment Manufacturers (OEM), and Solution Providers, which I will examine in this chapter. Indirect channels move product to the end user, and they generally offer more leverage—often a lot more leverage—than the field sales force channel. All have their economies of scale, and in today's competitive environment none can be ignored. Addressing any sales channel will still draw on all your skills as a seller, but you must also understand the "landscape" of sales channels that faces your company.

This chapter is an overview of that landscape, to give the seller perspective, and to provide an understanding of how to use channels effectively. Actually, making sales channels work has always been a critical success factor for technology suppliers. There are entire books on the subject, some of which are useful and should be read if channel sales is your primary focus.

First, some history: Not that long ago things in the channel were orderly, following rules that had existed for years. Products and services were passed from manufacturers to end users in carefully constructed buying chains (see **Figure 10-1**):

Figure 10-1: The Old Buying Chain

Then came the birth of Internet entrepreneurs who reinvented the distribution chain, skipping the intermediate layers, and causing havoc with how channels were used. **Figure 10-2** shows what the new buying chain looks like:

Figure 10-2: The New Buying Chain

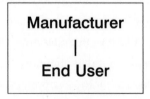

Nearly every manufacturer who can ship products that can be assembled easily by end users faces the temptation of using the Internet to reinvent the sales channel. In fact, new business models seem to be popping up every day. From where you are sitting, if enough different models are created, eventually the weak spots in your distribution model will come under fire. If this is where the margin lies in your value chain, the results can be devastating.

Some things do remain the same: Buyers must be reeled in quickly, and the sales channel remains the only way to connect a product with a market. And manufacturers of products and services still have two primary responsibilities to the channels: to deliver a product or service that works as promised, and to create demand for it

But these primary responsibilities are also primary hurdles. All too often, a new product or service simply does not work as promised. Or, creating demand for it runs afoul of the fact that it offers benefits that are not easily understood. Marketing efforts could overcome these problems, but most firms do not have the money to devote to a large-scale marketing effort.

So the instinctive response is to send out salespeople to knock on doors. Since fielding a salesperson can cost $1,000 per day, given the salary, commissions, and overhead, a logical alternative is to try to achieve "leveraged sales" through channels of distribution. That means sending the salespeople to knock on the doors of organizations who field salespeople themselves, to get them to introduce your solution to their already existing buyer base. Additionally, companies are putting an increasing amount of energy on "how" to take products and services to the end user, bricks-and-mortar market instead of focusing on "what" to bring to the market.

In today's economy, the use of any one channel without the use of others that complement or compete against it will limit your company's market performance. For example, if you go to market using only a direct sales force, your cost of sales may be so high that it would make you uncompetitive. If you use the Internet as your only sales channel, you sacrifice buyer service and loyalty.

No one channel can make a market. No one partner will do everything well and give you competitive advantage in and of itself.

To better understand the channels of opportunity, I'll divide the landscape into direct and indirect channels. Each channel offers its own set of advantages and challenges.

Direct Sales Channels

In direct sales, the product or service is sold directly to the end user. Fielding company-employed salespeople is the classic way to conduct direct sales, but there are others, and each may offer attractions. Here are a few to consider:

- Internet
- Telesales
- Government

Internet

Online ordering between corporations can be easier and more efficient than traditional methods, so more and more companies are turning to the Web to replace parts of the sales process. At the same time, many of these companies are keen to hold on to their salespeople, if only to push buyers toward the corporate site. In theory, the Web enables salespeople to make just as much money as before, by spending more time with buyers and earning new bonuses. The Web simply automates the routine parts of the job. But does that theory hold up in practice? Apparently it can.

Dell, Cisco Systems, and Intel, to name a few technology companies, do billions of dollars in business each year on the World Wide

Web and are successfully motivating buyers to order there. The trend is huge and growing ever more significant.

Dell, for instance, is known for being very serious about handling sales online. Dell currently does more than 50 percent of its total sales through the Net; that is billions of dollars in annual revenues. To keep salespeople from feeling threatened, the company pays bonuses when the staff meets quotas via online sales. They also pay salespeople to be in the field establishing relations with buyers, not at their desks waiting for a purchase order fax to come through. So Dell uses the Web as an ordering tool, not as a replacement for salespeople.

Dell's management feels that moving buyers online makes it easier to do business and to cuts costs. It also frees salespeople to focus on building relationships, forecasting, and finding new buyers, instead of wasting time on tasks such as order entry, purchase history, and pricing information.

Before the Web was around, many companies used product catalogs to show merchandise, and took orders by fax machine. Today, sales representatives can build home pages for individual buyers that have all the information online, along with ordering and configuration options, technical support and order tracking, and reporting features that help buyers keep track of where their money is going.

The flip side is that the Internet is troublesome for manufacturers, threatening to overturn the carefully nurtured network of distributors and Solution Providers they have used for decades. Simply stated the dilemma is this:

If you do not sell your products directly over the Internet, people will go to your competitors who do.

If you do sell your products directly, your distributors and dealers will desert you and only carry products from manufacturers who do not compete with them.

Manufacturers are dealing with this issue in a variety of ways.

Some display product information online without prices, and do not sell products directly over the Web. Rather, they help buyers find a dealer's location nearby. Others sell products directly over the Web but not at any discount, pointing buyers to the nearest dealer. Of late I have noticed a trend where vendors who sell products directly over the Web award commissions to salespeople or resellers for products sold in their territories. Still others sell products directly over the Web for what the market will bear.

Let's look at these strategies in detail.

Offer product information only, without prices or direct sales. Many companies offer only product information on their Web site, since they already have an extensive distribution system set up both in their own country and abroad. Consider that less than 5 percent of retail sales at present occur over the Internet. Would you be willing to endanger relationships with your distributors and dealers for that volume? For many manufacturers, the answer is a firm, "No, thank you." The risk is just too great.

Fear plays a big factor, too. Many corporate executives of established bricks-and-mortar companies are not yet computer-literate. They distrust and fear computer solutions, since they have become successful using traditional sales methods. In addition, they may have had complaints from their dealers about other dealers who are posting prices on the Web, lowering the street price and undercutting the competition. To protect their dealers from price gouging, they may prohibit all dealers from posting prices on the Web.

Sell directly without a discount. A second option is to sell products directly over the Web, but without any discount. This is an attractive option for many manufacturers. Lands End, a manufacturer of casual apparel, sells mainly through catalogs and retail stores directly to End User Stakeholders. But in order to increase awareness of their product over a wider audience, Lands End decided to sell directly on the Web, charging retail prices plus shipping to Internet buyers. This keeps them from competing with their stores for price but still gets the product out there.

Many publishers also use this approach. Both Simon & Schuster and John Wiley & Sons, book publishers, offer releases for sale on their site but at retail prices plus shipping and handling. This eliminates price competition with either traditional full-price retailers or online discount retailers.

Sell directly while paying commissions. IBM is experimenting with a number of forms of direct sales. Currently, it pays its sales force for all sales closed in their territories, whether or not the sale was made over the Web. The company is also developing a Team Player program where resellers register their buyers with IBM, who markets to them and then pays the resellers a fee whenever buyers make an online purchase. This form of direct sales is expensive and may only be useful for high-ticket/high mark-up items, but it does make for happy salespeople and resellers.

Sell directly at market prices. A final alternative is to sell at discount in direct competition with distributors and dealers. While I have yet to find any examples, I am sure many marketing vice presidents are tempted to give it a try. No matter how attractive the margins, however, this option is currently attractive only to very small manufacturers with little or no distribution chain to risk.

For now, manufacturers should balance the conflicting pressures of price competition, need for product awareness and accessibility, and maintenance of a traditional distribution network for the vast bulk of their sales. Many will choose to sell directly over the Web, but very carefully.

Telesales

Believe it or not, inside sales, or telesales, is the fastest growing segment for IT salespeople. I'm not talking about assembling an army of minimum-wage temps to troll the telephone books for unsuspecting victims. What I am talking about is a modern version that uses the Web as the seductive hook.

These days, anyone worth talking to has an account online. When you are talking to a prospect on the telephone, simply ask the prospect to call up a browser on the computer and click into a particular Web site. You, still on the phone, go to the same site. You then walk the prospect through a presentation or run a product demo. It is almost as good as being in the office with the prospect. The economics are compelling, too, with inside telesales people being able to "visit" many times more accounts in a day than any outside seller could ever get to. The trend I have noticed is for telemarketing operations to grow, and for outside salespeople to move inside.

Telesales has been found to be particularly effective for selling Internet services, such as those offered by Application Service Providers (ASP), which are companies providing Web-based access to software applications. These typically involve a modest monthly subscription fee rather than a steep onetime purchase price, keeping the barrier low. Also, such services lend themselves well to Web demonstrations.

Today's software vendors are very interested in making the Internet their primary sales and delivery vehicle, since it does away with the costs of product packaging and support logistics. Many of these products also lend themselves to online demonstrations.

As Web video cameras become ready for prime time, they will offer even greater advantages; the buyer will be able to see both you and what you are demonstrating by using an ordinary browser. It is said body language makes face-to-face interaction up to 50 percent more effective than using the telephone alone.

Of course, there are limitations to telesales. For the sales cycle to be short enough to be handled over the telephone, the solution must be well known or easily understood. It must have a low "barrier to entry," meaning the product must be both fairly cheap and not too complex.

Telesales is also not very good for building relationships, and face-to-face selling will still be required for complex, high-ticket sales. And you lose the chance to understand the politics of the buyer organization; you will not be walking down the hall with the buyer, trading stories, and being introduced to other executives in the elevator.

The Government

Surprise! The government, particularly the U.S. federal government, is a burgeoning sales channel. Oddly, many sellers make the mistake of considering the government channel as just another "sales territory." It isn't—it has its own distinct laws, buying cycles, and political landscape. But in most commercially driven organizations, government sales are a stepchild, often neglected and misunderstood by senior management. Looking at best practices in this area, companies such as Oracle, Sun, and Microsoft have proved to be effective at selling to the government. They all have these common characteristics:

- They recognize a separate "government" channel with its own profit and loss budget.
- The leader of this channel reports directly to the chief executive officer.
- The government channel has a separate legal counsel schooled in government procurement and practices.
- It has its own sales and support, based in the Washington, D.C., metropolitan area.
- It has access to a GSA (Government Services Administration) buying schedule.
- It has relationships with prime contractors to take part in existing procurement contracts.
- It offers products with a look and feel (including packaging,

documentation, language, and pricing) that says "government."
- It recognizes the political nature of the government and has a budget to hire government Subject Matter Experts as needed.

There are two big reasons for treating federal and state and local governments as a separate channel. The first is that the government is a potential pot of gold, much larger than any commercial account. And the government *will* spend billions each year on IT procurements or they will lose the budget for the next year. Because of this, most government sales are made in the government's fourth fiscal quarter (July–September) each year. Government agencies can show long-term loyalty, which can mean fortunes for its vendors. In the mid-1980s, for example, the government chose Novell's NetWare as its network operating system of choice. In the 1990s platforms such as NT and Linux replaced NetWare in the commercial markets, but many government agencies still use NetWare today, making these buyers cash cows for Novell. With this in mind, success in the government channel can be a life or death issue for a start-up as they grow into the mainstream.

The second reason is that the government channel is different from any other channel in that it is extremely "process oriented"; in other words, mired in red tape. It cannot be approached in the same manner as the commercial market. Securing the services of a government Subject Matter Expert, as described in Chapter 2, makes a lot of sense here. This person can also be your Champion Stakeholder. Government service is a brotherhood best understood by someone who has been there, or is there now.

Indirect Sales Channels

Over the years, sellers have swayed back and forth between direct strategies that are entirely or partly focused on end-user sales and channel-dominated strategies. Only a handful of vendors have actually succeeded by focusing on direct methods for the end user. Often, those that use the direct strategy return to the channel when the going gets tough.

Working with your resellers is a two-way street: Provide them with the right combination of elements for success and they will be committed to selling your solutions. So what do resellers expect manufacturers to provide?

- **A good technical support system.** When resellers know they can rely on your company's training, buyer service, technical support, and other branches of support within your company, they feel they can trust you. And trust is the basis of any solid relationship. If your company is available when resellers need you, it's easier for them to come to you with questions or problems whose answers may lead to a sale.
- **Good leads.** Solution Providers expect partners to provide them with leads, helping to create demand for the solution.
- **A good reputation.** If your company's reputation is questionable, resellers may have a hard time trying to overcome it in order to sell your products. A solid product, along with excellent service and a good public image, are all necessary to compete for today's media-savvy buyers.
- **A good product.** Even if a few poor-quality products slip out your door, it will be hard for resellers to help you out. Not only will buyers think twice before buying your product again, but also they will tell other prospective buyers about their experience. Besides, most resellers will plainly refuse to represent you if you are known for poor quality.
- **Good margins.** Resellers are concerned with the bottom line. They need to meet their expenses, just as you do. If resellers

know they are not going to profit much from a sale, they're not likely to go out of their way to sell your product. They will turn their attention to other products with bigger margins.

- **Good salespeople.** Your salespeople in the field are the buyer service representatives, marketing managers, and general face of your company in the eyes of your resellers. If you can provide your resellers with salespeople who are on top of the market and know the game, you have given your resellers a real advantage and a source they can rely on to take in with them to stakeholders.

Resellers may not tick off each item on this list of what they are looking for in representing a manufacturer. But when all of these are in place, they will be in a superior position to better serve your company's needs.

Let's look at specific indirect sales channels used in the IT arena.

- Solution Providers
- Distribution
- Service Providers
- Retail
- Original Equipment Manufacturers (OEM)

Solution Providers, Value-Added Resellers (VAR), and IT Consultants

In the past few years, the lines between Solution Providers (those resellers who provide software, hardware, and services as a bundled business solution to a buyer) and Value-Added Resellers (VARs) and IT consulting organizations have become blurred, practically beyond recognition. The important point, though, is that there are more lines to blur. New business models are emerging every day as old ones morph, collide, and disintegrate, and it can all be confusing to the end user. What hasn't changed is their focus on stakeholder-needs analysis, integration of hardware and software, and maintenance and upgrades, which are the Solution Providers' key services. This has long been their focus in various forms.

Typically, they specialize in niche markets, such as telecommunications, healthcare, financial, and engineering. They often like to sell complete packages to users in their niche, sometimes relabeling everything with their name. If you can interest them in including your product in that package, they essentially become part of your sales staff. If you sell software, they may combine it with hardware. If you sell hardware, they may combine with software and peripherals. Or their added value may simply be expertise in their niche, and they may sell your product as is. Many times, they are given a volume discount in exchange for a minimum yearly commitment.

The changing market is putting resellers in a unique position to win more sales, yet few resellers fully realize their potential advantage: the critical ability to sell technology-based solutions to senior executive buyers. And, if single-product Solution Providers don't move fast to exploit this advantage, they may find themselves squeezed out of the marketplace by systems integrators and other channel distribution alternatives willing to serve the same niche markets by integrating diverse products into equivalent solutions.

Developing lasting trust and credibility with high-ranking stakeholders has always been necessary for high-tech sales success, but recent research by Computer Reseller News (CRN) indicates that this talent is rising in importance. Two factors drive this emerging trend: the growing demand for fully integrated technology solutions, and the resulting desire by buyers to minimize risk by forging enduring partnerships with strategic suppliers.

Industry analysts, including the Gartner Group, find the ability to provide cost-effective integrated solutions is becoming essential to the survival of Solution Providers. We can thank the rise of the Internet and the growing demand for computing for this change. To realize new Web-based solutions for electronic commerce, knowledge management, and partner relationship management, to name but a few, buyers require a mix of hardware, software, and services—usually from multiple suppliers working in concert.

Recently, there has been a massive turnover among Solution Provider executives responsible for managing channel programs. These firms made significant changes to their business models as the result of the maturing of Internet technology. In an era of e-business and the new economy, however, it is still the basics that get the job done. The majority of Solution Providers offer needs-analysis, integration, network design, training, and hardware maintenance/upgrades as their primary services. Moreover, popular services such as local area network/Internet connectivity, e-mail integration, security, technical support, and Web site development will be the top ones provided by these end-user advocates in the coming years, and many will offer Web hosting as a key service to their buyers.

Solution Providers are the scrappiest bunch of business people out there, constantly forced to adjust to the twists and turns of the technology industry. Despite all the challenges they have faced over the years, theirs will continue to be a healthy business.

Distribution

While Solution Providers represent a one-tier channel, distribution is a two-tier channel: You sell to distributors, who sell the product to hundreds or thousands of resellers. With the distribution channel comes a pure division of labor. The manufacturer creates a product that delivers what it promises, and generates demand for that product. The distributor handles warehousing, collection/credit, and other logistical issues.

Distributors can increase market coverage, but can also distance the manufacturer further from the end user. This is a significant issue for start-ups, since you need maximum communication with the buyers in order to respond to their reactions to the product. You can respond by creating user focus groups and soliciting much-needed up-to-the-minute user feedback.

The big question is: Do you have a product that fits well in the distribution channel? Distributors are extremely effective at moving well-established, easy-to-understand products to a mainstream market. So if your solution is in a box or is a commodity, or at least something that can be neatly described as an SKU (Stock Keeping Unit), the better it will be for distribution to add value. Big-ticket or customized items do not fit well in this channel.

Distributors will also want to know whether carrying your product is worth their time. The larger distributors, Tech Data and Ingram Micro, will want to see sales of at least $2 million in the first year for a start-up and substantially more for an established company. They will then give your product ninety days to demonstrate a rate of sales projected to meet initial quotas. I have seen some vendors given more leeway if they decide your product is "strategic"—but don't count on it.

The Service Provider Phenomenon

In recent years, the IT marketplace has defined a new channel category. These new resellers provision software applications via the Internet and are aptly called "Service Providers."

At first, they were called them Application Service Providers (ASP). There is much anticipation in the probability of rapid growth of the ASP market. However, the ASP term itself has suffered the same fate as the dot-com term that came and went before it. In an attempt to distance themselves from the dot-com disaster, ASPs have adopted new terminology to buy themselves some time to reach profitability. ASPs have lost some steam, and so may the acronym that got the ball rolling.

More recently a new term "x" Service Providers (xSP) is used to describe a broad range of service providers offering many different types of services using the Internet. These services add support for business processes, are based on a one-to-many approach and priced on a recurring service fee or subscription basis.

While describing the xSP is relatively painless, not all xSPs are the same. Since the onset of the term *ASP,* companies have been redefining the acronym to better explain their core competencies. The result has been a landslide of acronyms such as Internet Service Provider (ISP), Financial Service Provider (FSP), Vertical Service Provider (VSP), and Managed Service Provider (MSP), which has created an overwhelming confusion in the marketplace. The newer term, xSP, which is a catchall service provider acronym in IT lingo, may be here to stay.

Currently, mainstream bricks-and-mortar companies such as WorldCom, Marriott, and American Express are providing these Internet-based services to their enterprises as well as their customer base. As companies understand the applications and adapt this emerging technology as a reliable delivery method this new channel will expand exponentially. Outsourcing software and services continue to grow and will fuel the need for Service Providers now and in the future.

Retail

With the rise of the Internet, this channel has been hammered hard by end users. Retail stores such as CompUSA, MicroLand, and Computer City provide easy access for the end user but many are now using the Web as a "storefront." Users view the product at the bricks-and-mortar retail store and then buy it online at a significant discount. This is forcing some retailers to change their business model.

Gateway, for example, established retail storefronts where customers view the merchandise and also order online. Apple Computer is in the process of implementing a similar retail strategy.

If you believe your product fits the retail channel, your first move should be to approach the distribution channel. Distributors can handle the chains and stores and relieve you of the paperwork.

Your second move should be to approach online retail sites and look for volume purchase commitments. Don't be surprised when online retailers are picky about the list price; people are often more willing to pay for a big-ticket item at a bricks-and-mortar store, where they can talk with a salesperson and test the product in advance. They cannot get that kind of reassurance online, and as the price goes up they are less willing to hand over their credit card numbers.

Expect, too, continued change. Those who joined the business-to-consumer e-commerce gold rush in the late 1990s are beginning to trudge back to civilization. Venture capitalists thought such sites could scale into major powers much faster than was humanly possible. As in the cases of Value America, eToys, and Pets.com, the management teams they assembled at breakneck speed lacked the necessary business-to-consumer experience. Even if sales were good, their business models demanded that management be extraordinary. When they weren't, investors turned away.

As various dot.com dramas play themselves out, traditional retailers have been slowly but effectively moving to establish online presences, and can eventually be expected to carry as much weight online as offline. But that could leave even more bricks-and-mortar stores struggling to justify their existence, as people learn to trust the Internet, and find that buying all their Christmas gifts online, at one sitting, is the way to go.

Only time will tell how it will all sort itself out.

Original Equipment Manufacturers (OEM)

Original Equipment Manufactures (OEMs) are those companies who bundle products and services and resell them to end users under a "brand" name. For example, PC manufacturers such as Dell, HP, and IBM, use OEM Windows software from Microsoft and distribute it with their PCs to buyers. Sometimes OEMs are somewhat misnamed, since it is often impossible to say who the original manufacturer of a piece of equipment or software is. Specialty firms manufacture the various components and the assembler slaps a brand name on the finished item and sells it through one or more channels. The company who owns the relationship with the end user is considered the OEM. Therefore, "OEMing" can broaden your market coverage but diminish your relationship with the end user.

Your product does not have to be a metal housing or an electric motor to fit in the OEM channel: A considerable amount of third-party software and finished peripherals go into end-user items. If you do find yourself approaching the OEM channel, keep in mind that the sale may hinge on the technical specifications of the product. It is also a business sale featuring deep discounts that can exceed 60 percent of the published price. On the positive side, you are freed from packaging and marketing the item; you just hand it over in raw form. If it's software, for example, you are selling a license to access the source code.

But if you like a quick turnaround, the OEM channel should not be your first choice, since the sales process is typically lengthy and more strategic in nature. After all, your product has to be designed into another product. Also, the buyer may want assurances that you can meet ongoing quality standards, and that you will be around when buyer service issues come up.

Summary

As time to market was the key in yesterday's technology markets, the key in today's markets is rapid and organized channel deployment. Your channel partner mix must change faster than it ever has before, and a proper mix—bringing the right users together with the right products—is the only way to achieve high-growth, sustainable, and competitive channel advantage.

Before you go out and practice some of the techniques you have learned in these pages, remember this: IT selling is about control of the process. The question is: Who is in control, the buyer or the seller? *I.T. Sales Boot camp* gives control to both. The buyer is in control because the seller is not pressuring the buyer to do something that the buyer is not ready or willing to do. And the salesperson is in control because the salesperson can help the buyer make decisions based on stated wants, facilitating the buying process in a business-driven yet emotionally charged process. The selling team is in control because sales management has the necessary tools to forecast business.

Everyone wins.

Appendices

Sources of Information

No single source of information can meet all your needs, but this collection of Web sites should get you started, especially in fields related to IT selling. I have arranged the sites by topic:

- Professional sales portals—for general and career information
- Business research sites—for digging into corporate backgrounds
- Sites offering mailing lists—for uncovering leads
- Sites for help with writing proposals
- Sites for help with creating presentations
- Web conferencing sites—for conducting "virtual" meetings
- Customer Relationship Management (CRM) sites—for information, software, and services
- Office organization sites—to help organize all components of an office via the Web
- Road-warrior sites—for mobile professionals
- Map-generating sites—for those same peripatetic professionals
- Sites that assist in the channel management

Professional Sales Portals

www.ITselling.com—This is my Web site. I believe it to be the complete online toolbox for the IT sales professional, with sales training articles and IT industry news, plus information on sales training seminars, online sales force automation tools, and Web conferencing.

www.saleslinks.com—Mentor Associates Inc. maintains this exhaustive listing of sites of interest to sales professionals, offering techniques, career information, personal development, sales technology, discussion groups, and inspirational material.

www.strategicaccounts.org—The Strategic Account Management Association (SAMA) is an international, nonprofit organization devoted to developing and promoting the concept of customer-supplier partnering.

www.justsell.com—This online magazine and discussion forum for sales professionals selects a different topic each month (such as "phases of the sales cycle" and "selling as a career") and offers related articles and software tools.

www.sales.about.com—Here you can find ongoing series of articles by and about sales professionals.

www.sellingpower.com—Maintained by *Selling Power* magazine, this portal includes links covering many topics of interest to sales professionals.

www.smartbiz.com—Here is an online encyclopedia of business-related articles, with many focusing on sales and selling.

www.amanet.net—The American Management Association's site offers a wealth of generic sales training material.

www.prosales.com—The place to go for news and information about sales and for sales-related services, such as presentation development.

www.salesrepcentral.com—Another sales portal.

Business Research

www.hoovers.com—This company has both a free and a subscription-based Web site, offering corporate information—including financials and names of executives—on thousands of public and private firms. The subscription service offers in-depth information; the free service will suffice for corporate phone numbers and addresses.

www.excite.work.com—The business information portal boasts business and industry news, articles on business-related topics, links to sources of business, and financial news. It claims to have profiles on more than 11 million firms.

www.edgar-online.com—Want to read a company's annual report? You can do that here, and search for filings with the Securities and Exchange Commission.

www.prnewswire.com—Press releases issued by corporations are usually posted here, where you can follow the news and search archives. (The same press releases are often posted on the corporation's Web site as well, but at *www.prnewswire.com* you can compare them to the competition's releases.)

www.individual.com—This division of *www.office.com* provides individualized news via the Web, free of charge, delivered in a morning e-mail.

www.office.com—This news site and discussion forum follows news and trends in 150 different market niches.

www.findsvp.com—This site offers access to live consultants for quick advice and research. Plus, you can contract for customized research, such as customer satisfaction surveys. This is not a free site.

www.census.gov—The Census Bureau does more than count heads, and the result is a lot of business and demographic data, which you can find here.

Leads

www.usadata.com—Specify and order mailing lists and purchase a wide variety of market research reports and demographic profiles.

www.infousa.com—Specify and order mailing lists drawn from files of more than 12 million American businesses and 161 million households.

www.dnb.com—At Dun and Bradstreet's Web site, look for business credit reports and acquire various business analysis tools.

www.harrisinfo.com—The Harris InfoSource site boasts a database covering more than 400,000 manufacturing establishments.

www.superpages.gte.net—This "phonebook" site lets you search for businesses nationwide, by name and telephone number.

www.cyberdirect.com—Cyberdirect focuses on the direct marketing business and features links to businesses that offer direct marketing tools, including mailing list services.

www.postmasterdirect.com—PosterMasterDirect offers e-mail mailing lists, including e-mail addresses of people who have "opted-in" at one of its corresponding e-commerce sites. (In other words, the people on the lists at some point indicated a willingness to receive e-mail promotions. E-mail sent in bulk to people who have not opted-in is considered spam.)

www.leads-link.com—Here is another site offering access to opt-in e-mail lists.

www.justleads.com—This is a branch of justsell.com that offers free access to its database of companies.

www.manufacturersnews.com—This online home of *Manufacturers' News* provides access to a database of U.S. manufacturers and executives.

www.myprospects.com—Search and download marketing lists based on predefined profiles that you select.

www.thomasregister.com—You may have seen the fourteen-volume Thomas Register in the library. Now you can access its contents at this Web site, for free.

www.zapdata.com—This site sells lists, company profiles, and industry reports from various sources.

www.bidradar.com—Here is where you can see what RFPs are being issued by the U.S. federal government, using material from the official *Commerce Business Daily*.

www.daily-leads.com—Get a steady supply of qualified leads, generated when people opt-in to any of daily-leads.com's network of business-related Web sites.

www.findmorebuyers.com—This service offers customized mailing lists in a number of consumer and business niche categories.

www.americanmanufacturers.com—The International Manufacturers Gateway includes a tool for analyzing market chains, plus an extensive database of manufacturers.

www.trueadvantage.com—This site contains information on relocating businesses, intelligence on the largest U.S. companies, intelligence on privately owned companies, new incorporations and start-ups, purchasing trends, emerging businesses, and RFPs/RFQs.

Proposals

www.ejustifyit.com—This is an excellent site that helps the IT sales team deliver a well-planned ROI (Return on Investment) to the buyer.

www.salesproposals.com—This site offers, for various prices, tips, tools, and books about designing and analyzing sales proposals.

www.pragmatech.com—Pragma Software offers proposal automation software; try out its "RFP Machine."

Presentations

www.spinitar.com—SPINITAR sells, integrates, and supports presentation systems, including communications and audiovisual units.

www.presentationuniversity.com—This site offers tutorials in effective presentations, plus software and clipart for PowerPoint and other formats.

www.presentationmaster.com—Here is a destination for news from the audiovisual industry plus hardware reviews and shopping guides.

www.presentersonline.com—This service produces graphic presentations, which you then access over the Web.

Web Conferencing

www.webex.com—This service lets you set up online meetings and presentations. Each participant needs a browser and a telephone; they connect via conference call and simultaneously look at the on-screen material, which can be annotated by each user. The service can also be used for software demonstrations.

www.placeware.com—PlaceWare similarly offers multi-seat Web conferencing with an emphasis on PowerPoint material.

www.blox.com—A merger of AlphaBlox and HalfBrain.com, Blox has software that lets you transform existing presentation files into Web presentations.

www.mshow.com—This service sets up Internet video presentations, handling everything from content development to after-show reports.

www.brainshark.com—This firm offers multimedia authoring software that, among other things, lets you add voicemail to a PowerPoint presentation.

Customer Relationship Management (CRM) Software

www.siebel.com—Probably the most comprehensive (and expensive) CRM solution.

www.crmassist.com—Click into news and information about CRM products and vendors.

www.crmcommunity.com—News, articles, and discussions about CRM and CRM vendors are on tap here.

www.crmguru.com—Find out what all the excitement is about concerning customer relationship management. The site is maintained by Front Line Solutions, a CRM consulting firm.

www.upshot.com—UpShot may be the best package available today for large account management. This customizable, end-to-end, sales management application lets you see into your sales pipeline and focus on the best deals.

www.salesforce.com—Salesforce offers very effective service to small sales teams. It boasts integrated customer support management, marketing reporting and analysis, best-of-breed sales force automation, and lead management.

www.agillion.com—Agillion emphasizes two-way collaboration and customer interaction, and provides a way to offer customers Internet access to customized account information.

www.salesnet.com—Salesnet is a feature-rich, user-friendly Web-based sales management application.

www.xsellsys.com—This service touts itself as offering mid-range CRM, along with sales training, skills development, and multi-channel and vertical market integration.

Organization

www.officeclip.com—This service lets you share documents according to your routing rules, and also share a calendar, bulletin boards, project tracking, CRM features, time management and expense reporting, paging, and other office automation features with members of your "virtual office." The service is free.

www.magicaldesk.com—Store your calendar, address book, message center, and files on this Web site.

www.interact.com—The company that sells the ACT! contact management software package also maintains this site, emphasizing the ways that ACT! can be put to use.

Portability

www.palm.com—The world is moving from your lap to your palm. In the United States this increasingly means PDAs (Personal Data Assistants) running the Palm operating system. You can find PDA information at this site and at its competitor, *www.handspring.com*. (PDA data entry is typically done on a desktop computer, so don't be put off by the tiny virtual keyboards.) Fitted with add-ons, PDAs also offer wireless communications and cellular connectivity.

www.palmgear.com—The advantage of Palm PDAs is that they can run applications, and you can find thousands of them here.

www.paypal.com—This electronic payment system includes a way to beam money from your PDA.

www.blackberry.net—An alternative to the Palm, the Blackberry PDA is increasingly popular, thanks to its real (albeit small) keyboard.

www.igo.com—The theme is one-stop shopping for accessories for mobile devices, as well as for notebooks, telephones, PDAs, camcorders, two-way radios, and voice recorders.

www.point.com—If you are interested in getting a cell phone, this site will let you research the service plans and phones available in your city.

Map Services

www.mapquest.com—There is nothing like a customized map that shows the precise location of your destination, so that getting lost on a sales call becomes one less thing to worry about. You can even download the maps to a PDA; driving directions and a road trip planner are available as well.

www.mapblast.com—This service is similar to Mapquest, except it will also add fast-food restaurants to your map.

www.travroute.com—This site offers products to help with complicated travel planning or to map out sales territories.

www.garmin.com—This is the place to click into if you are interested in consumer electronics that let you get around using GPS (Global Positioning System) technology.

Channel Selling Tools

www.partnerware.com—Partnerware offers Partner Relationship Management (PRM) software to manage indirect channel sales.

www.channelwave.com—PRM software at this site focuses on the "Partner Loyalty System."

Glossary of IT Selling Terms

Amiable type: Hardworking, team-oriented, somewhat rigid person who is often found at the End User, Champion, Cost Evaluator, and Technical Evaluator stakeholder levels (see Chapter 3).

Analytical type: Numbers- and detailed-oriented person, often lacking in social skills, sometimes found among Cost Evaluator and Technical Evaluator stakeholders (see Chapter 3).

Bad sales: Sales that turn out to be not in the interests of the buyer.

Behavioral styles: Observable behavior patterns among buyers that affect (along with their stakeholder position) how they should be approached. Styles include Take-Charge, Optimist, Amiable, and Analytical types (see Chapter 3).

Benchmarking: The practice of finding out, through legal methods, how another company does something better than you do, and learning from that company.

Bricks-and-mortar: Company whose business is conducted through traditional means, as opposed to a dot.com company.

Business line expert: Member of the sales team brought in to provide insight and introductions concerning a specific niche market. This person is usually an outside consultant, often a retired executive from the niche industry in question.

Champion stakeholder: A person (or persons) interested in seeing your sale succeed, and in a position to help move it forward. The Champion may be inside or outside the buying organization.

Competitive intelligence (CI): Information about competitors and their products, especially information about how their products and prices compare to yours.

Competitor: Any company that is capable of meeting your buyers' needs as well as you can or better. A "competitor" is no longer just a fellow player in your industry who provides products and services similar to yours and prospects the same types of buyers.

Complex: Where groups of people affect the outcome of a sale.

Cost Evaluator stakeholder: Person in the buying organization whose job it is to evaluate a proposed solution on price versus other vendors, and who selects a vendor of choice.

CRM: Customer Relationship Management.

CXO: Executive at the level of chief executive officer, chief information officer, chief financial officer, chief operating officer, and similar positions (that is, only the middle initial is different).

Direct sales: The selling of a product directly to End Users.

Dot.com: A company whose business is carried out primarily over the Internet, as opposed to bricks-and-mortar companies.

End User stakeholder: Person in a buying organization who will use the solution being offered by the seller.

Executive staff: Sales management and top management of a selling firm, forming part of the sales team.

Financial stakeholder: Person who has authority in a buying organization to move on a sale and to have a check written.

GAP model: Method the seller can use to pose questions to the buyer to elicit implied wants and turn them into stated wants (see Chapter 6). The questions should be asked in this order: Gathering, Association, Problem-Solving.

Gatekeeper: Person who answers the CXO's telephone and decides if you should be allowed to talk to the CXO.

Government channel: Sales channel embodied by government agencies as customers. Their size and preoccupation with process make them a separate sales channel.

Graphics designer: Member of a sales team who prepares visual elements for inclusion in proposals or presentations.

Implied wants: Wants that the buyer has no emotional attachment to but could be led to make such an attachment via the GAP process (see Chapter 6).

Indirect sales: The use of middlemen, such as distributors and retailers, to reach end users.

ISP (Internet Service Provider): Firms that provide Internet access and especially Internet presence to organizations and the public.

OEM (Original Equipment Manufacturer): A seller who offers products under its brand name that are actually assembled from the products of other manufacturers.

Optimist type: Relationship-oriented, noncompetitive people often found among Technical Evaluator and Champion stakeholders (see Chapter 3).

Program manager: Another name for the Proposal Leader of a sales team.

Project manager: Another name for the Proposal Leader of a sales team.

Proposal leader: The member of the sales team who writes the sales proposal, using information from the rest of the team, and who often coordinates the efforts of the other team members. This person may also be called the Project Manager or Program Manager.

RFP (Request for Proposal): Document issued by the buyer inviting sellers to propose a solution, usually in the form of a written proposal.

Sales channel: Chain of commerce between the manufacturer and the end user. It may be multi-tiered, as with wholesalers, distributors, and retailers, or single-tiered, as with direct Internet sales.

SECRET values: Topics of universal interest in the information technology field that can be used to create value by the seller: Security, Expandability, Cost, Reliability, Ease of Use, and Throughput.

Solution Providers: Selling organizations that offer specific solutions often using products from multiple vendors.

Stakeholder: Person in a buyer organization who is involved in influencing the buying decision. Types of stakeholders include Financial, Technical Evaluator, Cost Evaluator, Champion, and End User.

Stated wants: Wants that the buyer has an emotional attachment to and that will motivate him or her to move on the purchase (see Chapter 6).

SWOT competition analysis: A form used to analyze the strengths, weaknesses, opportunities, and threats associated with an account or opportunity.

Take-Charge type: Decisive, action-oriented individuals, often found among Financial Stakeholders (see Chapter 3).

Technical Evaluator stakeholder: Person in a buying organization (usually in a staff position) who reviews new technology and refers the winner to the Financial Stakeholder.

Technical expert: Sales team member with product knowledge extensive enough to back up the sales representative. The expert may be a product manager, systems engineer, or person in the firm, such as the founder.

Technological competitor: Firm that consciously uses technology as its primary competitive weapon to create new markets or to gain access to widely different markets. A primary aim of such competitors is to replace existing competition using leveraged, technology-based efficiencies of scale.

Telesales: The use of phone conversations to sell, often while showing the buyer material on the Web.

Value: The concept that the seller must establish for the buyer, while causing the buyer to want to buy.

Wants: Expressions through which buyers make final decisions on a purchase. They do not necessarily define what the buyer actually needs. *See* Implied wants and Stated wants.

White Paper report: Document submitted to the Evaluator Stakeholder of a buyer before an RFP is issued, in the hope of having its viewpoint reflected in the RFP.

Glossary of Information Technology Terms

Access Provider: A company that provides Internet access.

ADS: Asymmetric Digital Subscriber Line, a form of DSL.

AI: Advanced Interactive Executive, IBM's version of Unix.

Aliasing: Visibly jagged steps along angled or object edges, due to sharp tonal contrasts between pixels.

Alpha test: The testing of software internally, by the vendor's staff, as opposed to beta testing.

Analog: Data carried by a continuously variable signal or data, as in an old-fashioned telephone line.

ANSI: American National Standards Institute, the organization that establishes U.S. technical standards in many areas, including computers and communications.

API: Application Programming Interface, used by programmers to write applications to run under a certain operating system or with a certain umbrella application.

Application: Software that lets users do specific, complex tasks, such as word processing, database management, or accounting, as opposed to operating systems that control the basic functions of the computer.

ASCII: American Standard Codefor Information Interchange, a standard character-to-number encoding scheme used throughout the computer industry—except on IBM mainframes.

ASIC: Application Specific Integrated Circuit, a custom-made microchip.

ASP: Application Service Provider, a company that offers the use of software as a subscription-based service using the Internet as its delivery method.

ATM: Asynchronous Transfer Mode, a high-speed transmission method for data.

Bandwidth: Transmission capacity of a connection stated in bits per second. Local area network bandwidths start at ten megabits with Ethernet coaxial cable, while optical trunk lines are in gigabit ranges.

Baud: Bandwidth of a line measured in the number of signals that can be introduced into the line per second. The term formerly was used as a synonym for bits per second, but is obsolete for that purpose since modern equipment squeezes multiple bits out of each signal.

BBS: Bulletin Board System, a dial-up service often running on a stand-alone computer, maintained as a hobby or for business communications and now largely supplanted by Internet e-mail and Usenet.

Beta testing: The field testing of software by selected (or initial) customers, as opposed to alpha testing, which is conducted internally prior to beta testing.

Binary: Numbering system with only two values, zero and one. Binary is the basis of computer logic.

BIOS: Basic Input-Output System, a rudimentary operating system built into a machine, so it can load the rest of the operating system when it is turned on.

Bit: Single on-off signal or storage unit, representing a value of one or zero.

Booting: Starting a computer by switching it on, as opposed to rebooting, which is done after it is already on.

Broadband: Usually refers to data transmission at a speed faster than the 56,000 bits per second available with a modem, such as DSL, cable TV modems, and T1 lines.

Browser: A program that facilitates access to the World Wide Web.

Bug: Mistake encountered in a computer program.

Byte: One character of information, usually eight bits wide.

Cache: Segment of RAM reserved for data recently read from the hard disk, to keep it on hand for faster processing.

CAP: Competitive Access Provider. *See* CLEC.

CDR: Call Detail Recorder, a system that logs the destination and duration of telephone calls but does not necessarily record their contents.

CD-ROM: Compact Disc, Read-Only Memory, a disc that looks like an audio CD but carries data in computer format, typically more than 600 megabytes, and often used for software distribution. Standard CD-ROM drives are read-only, but some are read-write.

CGI: Common Gateway Interface, a standard way of running a program on a Web server so that applications can be attached to Web pages.

CLEC: Competitive Local Exchange Carrier, a newcomer telecommunication company that competes with an ILEC, sometimes using lines leased from the ILEC.

Client/server: A system whereby part of an application runs on a desktop computer and the rest runs on a server elsewhere on the network, one responding to requests from the other. The one making the request is the client.

CMIP: Common Management Information Protocol. An industry-standard protocol for accessing and controlling network components.

CO: Central Office, the building where local phone lines are concentrated onto a backbone network.

Coaxial cable: Cable with a central insulated conductor, wrapped by a braided-metal conductor and an outer layer of insulation.

Compression: The act of reducing file size, typically for easier transmission, by eliminating redundancy. The average file length can be cut in half, but results vary greatly, especially with graphics files.

Cookies: File sent to a Web browser by a Web server that is used to record the user's activities at a Web site.

CORBA: Common Object Request Broker Architecture. An industry standard method for creating, distributing, and managing "program objects" (i.e., software programs) throughout a local network.

CoS: Class of Service, as in how many options you have on your phone line.

CPE: Customer Premises Equipment, as in the telephone you own.

CPU: Central Processing Unit, the brains of the computer, typically embodied in a microprocessor chip. All other components of the computer exist to feed data and instructions into the CPU, and to retrieve, display, or store the results.

Crash: A condition where the computer has unexpectedly stopped working, typically as a result of a software bug or hardware malfunction.

CRC: Cyclic Redundancy Check. Extra data bits added to a data transmission that can be used to check for the possible presence of errors in the rest of the transmission.

Cross-platform: Software (or hardware) that will work on more than one type of computer.

Cyberspace: Everything available and happening online, such as Web pages, chat rooms, e-mail, download sites, and private bulletin boards.

Database: A file that contains a collection of information organized into records, each of which is further divided into fields.

Decompression: The expansion of compressed files to their original state and size.

Demo software: Version of a software package that lacks the full features of the original package or that cannot be used beyond a specific date, used to demonstrate the capabilities of the original full-function package.

Digital data: Material encoded into discrete on-off signals, as opposed to analog signals.

DIP switches: Dual Interface Poll switches, small on-off settings found on circuit boards that allow the user to customize their configuration.

DLC: Digital Loop Carrier. A circuit box, often by the roadside, between a subscriber and the phone company central office, which can be used for, among other things, extending the reach of DSL.

Domain name server: A computer that converts human-readable Web or network addresses, such as *www.yahoo.com*, to a numerical IP address that the network uses internally.

DOS: Disk Operating System, the original operating system used on many personal computers. Originally, Windows was simply an interface that ran on DOS.

Download: To retrieve a file from another computer using a modem or an Internet connection. The opposite is "upload."

Driver: Software that tells the computer how to operate an external device, such as a printer.

DSL: Digital Subscriber Line, broadband service over ordinary telephone lines.

DSLAM: Digital Subscriber Line Access Multiplexer, used to connect a DSL line to a central office backbone.

Duplex: The direction of signals on a transmission connection. Full duplex means data are flowing in both directions at a time. Half duplex means data can flow in either direction, but only in one direction at a time. Simplex means data only flow in one direction.

DWDM: Dense Wavelength Division Multiplexing. A signaling method that increases the capacity of the fiber-optic cables used in telephone company backbone networks.

EBCDIC: Extended Binary-Coded Decimal Interchange Code, the character code used by the few computers that do not use ASCII.

EF&I: Engineering, Furnishing, and Installation. The act of planning the workings of, acquiring the equipment for, and installing the equipment in an IT facility.

E-mail: Electronic mail, text messages sent between users over a network. The Internet allows file attachments, but some older corporate e-mail systems do not.

Encryption: Methods that turn files and messages into what appear to be random numbers so they can only be read by the authorized recipient with a decryption key. Encryption is also used to authenticate the identity of the sender.

Ethernet: Common LAN protocol typically running at ten or 100 megabits.

Extranet: An intranet that can be reached from outside the organization, through dial-up telephone lines or by password-protected Internet access.

FAQ: Frequently Asked Questions, a document that functions as a general introduction to a topic using a question and answer format.

FCC: Federal Communications Commission. The government agency that regulates the broadcasting and telecommunications industries in the United States.

FDDI: Fibre Distributed Data Interface. A method of sending data over high-speed fiber-optic networks.

File: Stored information on a computer disk that can be retrieved under a single name. The formatting and nature of the data in the file depends entirely on the software that stored it.

File server: Computer that stores files for use by other computers on a network.

Firewall: Computer program (or dedicated computer) that isolates a local network from the Internet, permitting only specific traffic to pass in and out.

Flaming: Hostile reactions, delivered in e-mail messages, to newsgroup postings or e-mail. Often results from misinterpretation of an e-mail message lacking the social cues available in face-to-face conversation.

Freeware: Software for which no payment is expected.

Gigabit: 10^9 bits of information, or about one billion bits of information.

Gigabyte: 10^9 (one billion) bytes (characters) of information.

Greenfield: A new ILEC using only the latest technology since it has no legacy systems.

GUI: Graphical User Interface, an interface such as Windows or the Macintosh OS with which the user trigger events by manipulating graphical objects on the screens. Non-graphic operating systems require the user to know and type in commands at a "ready" prompt on the screen, much as if he were sitting at a Teletype machine.

Handshaking: The process computers, modems, and fax machines go through with each other when initially connected.

Home page: The initial page displayed when a user reaches a Web address.

Host: The main computer in a network or the computer running the application that you are interested in using.

HTML: Hypertext Markup Language, the formatting system used to display files on browsers on the World Wide Web.

HTTP: Hypertext Transport Protocol. The protocol for moving Web files across the Internet. The servers that use HTTP, plus the desktop computers whose browser software display the transmitted files, comprise the subset of the Internet called the World Wide Web.

Hypertext: On-screen text that is linked to other text or material, calling it up when clicked or otherwise selected. The World Wide Web is based on the concept, since it addresses network navigation issues.

I/O: Input/Output. Refers to any of the ways used to get data into or out of a computer, including the hard disk, serial ports, printer ports, USB ports, modems, etc.

ILEC: Incumbent Local Exchange Carrier, a telecommunication company with cable in the ground, typically a member of the old Bell System.

Import: To bring data into an application, typically data generated by a different application.

Information Technology: The design, manufacturing, implementation, and use of computers and communications for the purposes of information processing.

Interface: The point of interaction between a computer and its user (as with some combination of the screen, keyboard, monitor, or joystick) or the manner in which a peripheral is attached to a computer.

Internet: Global system of interconnected computers sharing the same file addressing and retrieval protocol, allowing a user on one Internet host and access files on another other Internet host. The World Wide Web is a subset of the Internet, using hosts running the HTTP protocol. Other aspects include chat rooms, FTP sites, e-mail, Usenet, and numerous special services.

Intranet: LAN or WAN using Internet protocols, so that the same server software, browsers, and files used on the Internet will work on the LAN or WAN.

IP: Internet Protocol, the addressing scheme used internally by the Internet.

IRC: Internet Relay Chat, Internet facility that allows people around the world to chat by typing messages to each other. IRC is divided in thousands of topical chat rooms called channels.

ISDN: Integrated Services Digital Network, digital telephone lines that are slower than DSL or cable but faster than analog modems.

ISP: Internet Service Provider, a company that provides access to the Internet.

ISV: Independent Software Vendor. A company that specializes in the development and sale of software.

ITU: International Telecommunications Union. The international body that sets telecommunications standards so that, for instance, you can listen to radio broadcasts from another country.

IXC: Interexchange Carrier. Another term for a long-distance phone company.

Java: A programming language used to create small programs (applets) to download and execute as part of a Web page, although it has also been used for larger applications.

Kilobit: 10^3 (one thousand) bits of information, usually used to express a data transfer rate.

Kilobyte: A unit of data storage which represents 10^3 (one thousand) characters of information.

LAN: Local Area Network, a network of directly connected machines in the same office or building.

LDC: Long-Distance Carrier. A phone company that provides long-distance services but not local services.

Legacy: Any computer, transmission facility, software application, or stored data that is based on technology currently considered obsolete but remaining in use.

Linux: Operating system similar to Unix that is written and maintained by a committee and is available free, although "distribution disks" are sold and supported by various vendors.

LMDS: Local Multipoint Distribution System. High-speed wireless transmission systems based on carefully aimed rooftop antennas.

LNP: Local Number Portability. The ability of telephone subscribers under U.S. law to retain their local phone numbers if they switch to another local telephone service provider.

Lossy: Image compression that sacrifices some image resolution for greater compression ratios.

Mainframe: A large, multitasking, multiuser computer.

MAN: Metropolitan Area Network, a WAN serving a downtown area.

Megabit: 10^6 (one million) bits of information, usually used to express a data transfer rate per second.

Megabyte: A unit of storage representing about 10^6 (one million) characters of information.

Megahertz: A million cycles, often used to refer to CPU operating speeds.

Middleware: Software on a client-server network that passes data between the front-end software on the client (which formats and presents the data) and the backend software on the server (which accesses and processes the data).

MIPS: Millions of Instructions Per Second, a gauge of computer performance now fallen into disuse,

since there is no standard definition of "instruction."

Mirror site: Internet site carrying copies of material found at other sites, to ease network traffic conditions.

Modem: A device that allows computers to communicate over standard voice-grade telephone lines.

Moore's Law: The observation and prediction (rather than a physical law) that computer chips can be expected to double in power every eighteen months, at the same price levels. It has held true since the mid-1960s.

Motherboard: The main circuit board of a computer, often carrying the CPU, RAM, and I/O processing chips. Other circuitry may be carried on expansion cards and daughter boards.

MTBF: Mean Time Between Failures. The average time a system or component can be expected to operate until failing and requiring maintenance.

MTTR: Mean Time to Repair. The average time you can expect it to take to repair a system or component after its MTBF has been reached.

Multimedia: Computer material that involves more than text, including graphics, audio, or video material.

Multitasking: Capability of an operating system to appear to do several things at the same time, for one or more users. Actually, each task is given tiny time slices in succession.

Native: Software that is written specifically to run on a particular processor.

Netiquette: A form of online etiquette intended to avoid flaming.

Network: Any group of computers set up to readily communicate with one another, as in a LAN, WAN, MAN, intranet, or Internet.

Newsgroup: An individual discussion topic on Usenet, devoted to a specific topic.

NIC: Network Interface Card. An add-in board in a personal computer used to allow communications between computers. *See* LAN.

Object-oriented: Advanced form of programming in which both data and functions are used to define a procedure, which become self-contained "objects." Applications are then created by stringing together objects.

Operating system: Software that supervises and controls tasks on a computer. The application in turn uses the facilities of the operating system to access files, display information on the screen, and perform other tasks.

OSI: Open System Interconnection. A standard that lets networks from different vendors connect with each other.

Packet-switching: Data transmission process used on the Internet, whereby material is broken into packets for transmission and reassembled at the destination. No direct connect is necessary between the sender and the receiver. The opposite is circuit switching as with a telephone call, where a direct connection is maintained for as long as the call lasts.

PBX: Private Branch Exchange. The switchboard that connects the lines of a corporate phone network with the Public Switched Telephone Network.

Peer-to-peer: Networking protocol whereby each computer can request files and other services directly from each other without going through a central host.

Pixel: Individual dot that makes up a picture on a computer screen.

Plug-and-play: A facility in the later version of Microsoft Windows that allows the computer to reconfigure itself when new components are added.

POP: Point of Presence. Place where the backbone network of a CLEC, ILEC, or another telecommunications carrier is available for local connection.

POTS: Plain Old Telephone Service. If you have a phone that lets you dial someone outside your organization and talk to them, you have POTS.

PPP: Point-to-Point Protocol, used by dial-up Internet connections.

Protocols: Any process governing the way two devices exchange data.

PSTN: Public Switched Telephone Network. If you can pick up a phone and dial another phone that is outside your organization, then both your phone and the other phone is attached to the PSTN, which now reaches essentially the entire non-polar land surface of the globe except North Korea, and even there via satellite wireless links.

PTT: Post, Telephone, and Telegraph ministry, the European version of the telephone company.

QoS: Quality of Service. Refers to the nature of the attention that the network gives a particular connection. Voice calls require a higher QoS than data calls, since people don't like being cut off.

RAM: Random Access Memory, the kind of memory used internally in computers, using chips. The contents of RAM are lost when the system is powered down.

RBOC: Regional Bell Operating Company. A local phone company left over from the breakup of the Bell System in 1984.

Resolution: The sharpness of an image on a computer screen or a printed page, measured in dots per inch.

RGB: Red, green, and blue, the primary colors used by computer monitors. A given printer may use a different set of colors.

ROM: Read-Only Memory, used to store internal memory that must be kept permanently.

Router: Special-purpose computer used for traffic management on a network.

Search engine: Web site that can be used to search the contents of the Internet.

Server: Computer on a network whose main purpose is to share its resources (typically files) with other computers.

Shareware: Software that is distributed free of charge in the expectation that the user will pay for it if he finds it useful. The distribution version may have certain limiting features, such as the inability to save new files after a certain period of use.

SNR: Signal-to-Noise Ratio. The volume of useful information getting through a channel compared to the nonuseful information (noise) it contains.

SOHO: Small Office/Home Office.

SONET: Synchronous Optical Network, a high-speed protocol used by telephone company backbone networks.

Spam: Mass e-mail or newsgroup posting that tend to annoy the reader, often advertising get-rich-quick schemes that are little more than glorified Ponzi scams.

SQL: Structured Query Language, a data format used by many database systems to retrieve and modify information.

T1: A digital line carrying 1.544 megabits per second.

T3: A digital line carrying 44.736 megabits per second.

TCP/IP: Transmission Control Protocol/Internet Protocol, the transmission protocols used internally by the Internet.

Throughput: The amount of work that can be performed by a computer system or component in a given period of time

UNIX: An operating system that supports multiuser and multitasking operations favored by large Internet, academic, and research hosts.

Upload: To send a file to a distant computer using a modem or other transmission facility.

URL: Uniform Resource Locator, the addressing scheme used by the World Wide Web.

Usenet: A collection of online discussion topics, individually called newsgroups, associated with the Internet.

Virus: A program that replicates itself from one computer to another over a network, possibly corrupting system files in the process.

VoIP: Voice over Internet Protocol. A method for using the Internet to make free long distance phone calls, and also for moving voice calls over high-speed data lines.

VPN: Virtual Private Network. A private network created by using public Internet facilities, encrypted so that nonmembers cannot use them.

WAN: Wide Area Network, a network that spans an area larger than an individual building or campus, as opposed to a LAN.

Web Page: HTML document displayed on a browser screen, including the graphics associated with that file.

Workstation: High-end desktop computer typically used for engineering, scientific, or design purposes.

XML: Extensible Markup Language. A standard that allows users to create their own customized extensions of HTML for special purposes.

Glossary of Business and Financial Terms

10-K: Annual report in a format required by the Securities and Exchange Commission.

10-Q: Quarterly report in a format required by the Securities and Exchange Commission.

Analyst: Employee of a brokerage or fund management house who studies companies (usually in a specific industry) and makes buy and sell recommendations on stocks.

Announcement: The formal offering of a product to a target market. It may take the form of a simple appearance in a catalog or a press release, or be introduced in a full-scale, highly promoted and publicized "launch."

Annual report: Yearly record of a public company's financial position, including a description of operations, balance sheet, and income statement.

Arbitrage: To profit from the price of a security that is traded on more than one exchange.

Autoregressive: To use previous data to predict future trends. The fact that mathematical precision is possible does not mean the predictions are accurate.

Available market: The population of consumers or buyers who would be reasonably interested in a product or service, have access to it, and who have the money to buy it.

Back office: Clerical functions, such as payroll, that support the productive operations of the enterprise but that are not involved in carrying out those operations.

Balance sheet: A report of a corporation's financial condition at a specific time.

Barriers to entry: Considerations that make it difficult or expensive to enter a given market, such as well-entrenched competition and high start-up costs.

Benchmarking: Programs by which companies compare their business practices to those of other companies to see where their own practices can be improved.

Bundling: To offer multiple products and/or services as one product with one price. For instance, a personal computer often has Microsoft Windows and office software bundled with it.

Business cycles: Long-term patterns of alternating economic growth and decline, usually passing through periods of expansion, peak, contraction, and trough.

Catalog: A catalog company that creates the catalog, interfaces with customers, and takes and fills orders is considered a channel.

Catalog merchants: Vendors who produce catalogs (printed, online, or on CD ROM) as an efficient after-market sales vehicle to generate demand for niche products.

Channel: A group of companies who have a similar business model and are selling similar products and services to a similar population of customers.

Channel partner: A party that contracts with a vendor to act as a conduit for the vendor's products or services to a market. The role of the channel partner varies by customer type, supplier type, and business model of the various companies comprising the channel.

Chapter 11: Chapter of the federal bankruptcy code under which a firm continues operating while attempting to restructure its debts.

Chapter 7: Chapter of the federal bankruptcy code under which a firm closes down and liquidates its assets to settle its debts.

Cherry picking: To buy products individually from multiple vendors when they are commonly bundled by one vendor.

Cold-call: To make an unannounced sales pitch to a prospect, usually via the telephone.

Common stock: Stock that allows the owner to vote for officers of a company.

Competitive advantage: Characteristics of a firm that make it noticeably more competitive than other participants in its industry or market, such as higher-quality products, lower prices, or a brand name that is more widely recognized.

Computer dealer: Storefront operation selling computer products and services at retail. Also called a reseller, retailer, or aggregator.

Consolidation: The inclusion of subsidiaries in the financial statements of the parent company.

Corporate culture: Buzzword referring to the shared values, norms, and behaviors of a company, as reflected in its practices and products.

CPI: Consumer Price Index, a measure of price changes in consumer goods and services identifying periods of inflation or delation.

CPM: Cost Per Thousand, the cost, to a buyer of advertising, to reach 1,000 people with an advertisement. CPM is used to compare the price of advertising outlets in the same medium.

Differentiated marketing: To segment a market and offer products tailored for each segment. Also known as "niching the niche."

Direct (Distribution) channel: A channel that is owned by the vendor and thus acts as a direct conduit to the customer.

Direct marketing: Marketing initiatives, such as mail or telemarketing, aimed directly at members of a target audience, intended to generate sales rather than promote brand awareness.

Disclosure: The section of a company's financial statement that includes an explanation of its financial position and operating results.

Discretionary income: Amount of money someone has left after taxes and essentials.

Distribution channels: The routes through which products move from manufacturers to the ultimate consumers, referred to in a given market as "the channel."

Distributor: A wholesaler who also performs a certain level of integration, configuration, and testing. Also called a Master Reseller. *See* Chapter 10.

Dividend: Distribution of a portion of a company's earnings, cash flow, or capital to shareholders, in cash or additional stock.

Empowerment: Buzzword that, in the business world, refers to the practice of increasing employees' motivation by giving them more involvement in their work.

Events marketing: Sponsoring an event such as a car race in exchange for advertising during the event.

Fixed costs: Operating expenses such as rent and utilities that do not change even if there is an increase or decrease in sales.

Freelance: Work done on a per-job basis by an independent contractor, usually operating alone.

General availability: A product that can be stocked by channels and ordered by and shipped to customers without delay. IT products are often announced before general availability, potentially irritating customers who expect immediate delivery.

Global account manager: A business development salesperson whose responsibilities span national borders and who typically is concerned with market share rather than product issues. May play a role with customers who have global IT acquisition strategies, and may have profit and loss responsibility for the account worldwide.

Hybrid channel: A sales channel requiring multiple participants to take part in a single sales transaction with the customer. An example is a hardware vendor and software publisher teaming to provide a solution for a specific industry. Most IT channels are ultimately hybrids.

Indirect (Distribution) channel: Conduit to the customer that is not owned by the vendor of products that are being sold.

Inside sales representative: Sales representative who is expected to sell over the telephone. May be teamed with other representatives in an account or territory, or may be responsible for after-market sales. Also called a Telesales Representative.

Insider: Anyone who has access to material nonpublic information about a corporation, and including directors and officers as well as shareholders who own more than 10 percent of any class of equity security of a corporation.

Intellectual property: Copyrights, patents, and trademarks.

Internal rate of return: What's left of a rate of return after all payouts.

Internet channel: A direct channel established via the World Wide Web, acting as an electronic catalog sales organization.

Inventory turnover: The ratio of annual sales to inventory, used to gauge the effectiveness of a company's marketing efforts.

IPO: Initial Public Offering, the first sale of common stock by a corporation to the public.

ISV: Independent Software Vendor, a software company that sells vertical applications or specialty utilities, usually that it developed.

Latent market: A market for a product or service that does not yet exist.

License to distribute: A license between a software publisher and a channel partner that allows the partner to represent and distribute the publisher's products.

License to use: A contract between a customer and a software publisher that authorizes use of the publisher's product.

List broker: Someone who sells or rents lists, typically mailing lists, of people in specified target market or population.

M&A: Mergers and acquisitions.

Major account manager: A sales position dedicated to servicing sufficiently large corporate accounts. May have more than one major account unless the account is very large.

Market penetration: Percent of potential customers who have already bought the product in question.

Market share: Units sold by a vendor as a percentage of the units sold by all participants in that market.

Mass merchants: Retailers that carry large volumes of products at discounted prices. The number of brands offered is usually more limited than what would be found in a typical retail outlet.

Media buyer: Someone who buys advertising space or time in various media for a company that wishes to advertise, or for an advertising bureau.

Mission statement: Short statement by a company listing its strategy, purpose, and values. What is not included is usually more telling than what is included.

Net income: Profit, reflecting the cost of doing business, depreciation, interest, taxes, and other expenses.

Niche marketing: Serving a small, specialized segment of a larger market, presumably a segment that is of little interest to major competitors in that market.

OEM: Original Equipment Manufacturer, generic term for any company that licenses or purchases technology from another firm for use in a combined hardware or system solution that is marketed under the OEM's brand name.

Office equipment superstores: Office equipment store that moves high volumes of a limited number of brands at discounted prices. The model involves limited sales help, warehouse-style floor planning, and centralized distribution. Commissioned sales staff may handle the selling of computer equipment.

Overhead: Corporate operating costs such as utilities and maintenance that are not directly involved with producing products or services.

Partner: In terms of IT marketing, a company that forms a contractual relationship with another company that allows the partner to market a product to an end user.

PR: Public relations, an activity designed to increase public awareness and acceptance of a company and its products and practices, usually by promoting favorable coverage by news media.

Preferred stock: Stock that gives the owner first rights to distributed earnings but no right to voting.

Press release: Prepared statement distributed to various news media to announce corporate news. Unlike advertising, which is run as-is for a fee, press releases may or may not trigger news coverage, and the party who distributed the press release has no control over the content of the resulting media coverage.

Product mix: Collection of products that a company offers a market.

Professional services: In the IT arena, consulting services, education and training, software development, and applications management. Providers of such services generally have no ties to vendors in order to ensure credibility.

Profit margin: Earnings divided by sales, both measured over the same period.

Rate card: Brochure detailing the price of advertising in a particular media outlet.

Reseller: Generic term for any company authorized to sell or pass on a license for another company's products or services.

ROI: Return on investment, profit or loss resulting from an investment or purchase, often expressed as an annual percentage rate.

SEC: Securities and Exchange Commission, the primary federal government regulatory agency of the securities industry.

Security: As a business term, any paper that can be traded for value other than an insurance policy or a fixed annuity. The federal courts have ruled that if a person invests money in a common enterprise and is led to expect profits from the managerial efforts of the promoter or a third party, the investment is a security.

Shares: Certificates or bookkeeping entries representing ownership in a corporation or enterprise (i.e., where the investors stand to gain or lose by the success of the enterprise).

SIC: Standard Industrial Classification, each four-digit code representing a unique business activity or industry, administered by the U.S. government's Office of Management and Budget (OMB.)

Software publisher: Originator of software licensed under its name, although some or all of the software may have been written by third parties. The software may be sold to end users or incorporated into other products.

Strategic planning: Developing long-term goals and deciding on ways to meet those goals.

Super store: Similar to a mass merchandiser, except that it focuses on a narrower market, such as office products, computers, or shoes.

Synergy: Where the sum of combined efforts is more than the total sum of the efforts had they not been combined.

System integrator: Firm that supplies customized solutions using a project approach.

Target market: Market being actively pursued by a vendor.

Telemarketing: Direct market method whereby the consumer is called by the seller and invited to make a purchase.

Territory representative: Sales representative responsible for accounts in either a geographic or vertical industry territory, often with the help of a team of specialists. Generally has sales quotas and is expected to make face-to-face sales calls, although a considerable portion of sales may be over the telephone, especially to existing customers.

Transit advertising: Display advertising on vehicles, such as buses and subways, targeting specific geographical markets.

Underwriting: Guarantee of a bank to a company that is issuing securities that it will purchase unsold securities at a fixed price.

Unearned Revenue: Advance payment received for services not yet rendered or goods not yet delivered.

VAR: Value-Added Reseller, an OEM that sells solutions to vertical markets, often combining hardware, software, and services.

About IT Selling, LLC

IT SELLING, LLC is an international sales consulting firm based in Washington, D.C. Its clients include Sun Microsystems, NEC, Oracle, and dozens of other world-class sales organizations that have benefited from the processes described in this book.

It is the creator of the I.T. Sales Boot camp, a two-day classroom sales training program. Buyers can choose to implement this program in-house by having only their employees attend, or by sending participants to public seminars staged in major cities around the world. IT Selling currently offers the following seminars: IT Selling Boot Camp, Coaching IT Selling, and Presentation Skills for IT Selling.

If you would like to learn more about how IT Selling, LLC can help you and your organization, please visit our Web site at *www.ITselling.com*, or call us at 301-365-6600.

Index

A

account management programs, strategic, 149–154
    components of successful, 161–165
    creating partnerships, 167
    data, timeliness of, 153–154
    development and management responsibility, 157
    implementing, steps to, 166–167
    managers, authority of, 160–167
    managers, compensation for, 159–160
    managers, role of, 157–158
    organizational barriers, 162–163
    systematic approach to, 151–152
    which accounts to manage, 155–157

Acrobat Reader (Adobe), 125

active listening, 40–41

Aetna Life Insurance Company, 149–151

Amazon.com, 75–76, 138

amiable types
    buyers as, 41, 44–46
    defined, 199
    salespeople as, 51–52

analytical types
    buyers as, 41, 46–48
    defined, 199
    salespeople as, 52, 53

AOL (American Online), 83–84, 138

Apple Computer, 82

ASP (Applications Service Providers), 176, 183–184

association questions, 105–106, 110, 144–145

AT&T, 22

B

bad sales, defined, 199

behavioral styles of buyers, 40–41
    amiable types, 44–46, 51–52, 199
    analytical types, 41, 46–48, 52, 53, 199
    defined, 199

dominant (core), 48–49
financial stakeholders, 42, 43, 47
mixing, 48–53
observable, 39–40
optimist types, 41, 43–44, 50–51, 200
subcore, 49
take-charge types, 41–43, 50, 201
behavioral styles of salespeople, 39–40
benchmarking, 143, 199
best practices, 143
beta testing, 5–6
Bezos, Jeff, 76
Bosack, Leonard, 82–83
bricks-and-mortar companies, 136–139, 199
business line experts, defined, 199
Buy.com, 75
buyers
  as beta testers, 5–6
  emotional involvement of, 92, 94
  information on competitors, 141, 142
  questioning, 89–90, 92–104
  strategic, criteria for selecting, 154–156
  SWOT analysis of, 30–31
  technological expertise, x
  understanding, 39–40
  wants, 90–92
  why they don't buy, 38–39
  See also behavioral styles of buyers; Major Account Profile
buying chains, old vs. new, 170–172

C

Chambers, John, 83
champion stakeholders, 10–11, 14–15
  defined, 199
  gathering questions for, 97–98
  in government sales channels, 178

personality types, 43, 45–46
  referrals from, 56
chief (executive, financial, information, operating) officers. See CXOs
chief privacy officers (CPOs), 74–75
Cisco Systems, 83
cold calls, 56, 64–65
commissions, sales, 159–160, 175
Compaq computer, 136
competitive intelligence (CI), 136, 199
competitive readiness, assessing, 147–148
competitors
  and account management, 150
  collaborating with, 145–146
  defined, xii, 199
  identifying, 140–145
  knowing, importance of, 32, 135–136
  RFPs wired for, 115–117
  technological, strategies of, 139–140
  technological vs. bricks-and-mortar, 136–139
complex, defined, 199
computer chips, increasing power of, xii
computers, ease of use, 82–83
consultants
  IT, 180
  as subject matter experts, 26
core selling situations, 49
cost evaluator stakeholders, 10–11, 16–17
  defined, 199
  gathering questions for, 99
cost value, 73, 76–80, 86
CRM (customer relationship management)
  defined, 199
  software, 195–196
CXOs
  defined, 199

About the Author

BRIAN GIESE is an accomplished speaker, writer, and consultant in the field of information technology sales and sales management. Over the past two decades he has trained and motivated more than 200,000 salespeople with his warmth, candor, and skills-based processes.

As a sales and management executive with Novell, Brian became a member of the Society of Super Software Sellers and is the recipient of the Million Dollar Award for peak performance. He is an expert on the how-tos of building business relationships, and his clients read like a Who's Who in the IT sales world. He has served as a peak performance consultant for companies such as Sun Microsystems, Oracle, and NEC, as well as many Solution Providers of IT products and services. He is also a noted member of the National Speakers Association and is a frequent speaker at sales conventions and national sales events.

He is the president of IT Selling, LLC and the creator of the IT Selling Boot Camp used by many world-class technology sales organizations around the world.

Brian Giese can be reached at *brian@ITselling.com*.